LANGUAGE, KNOWLEDGE, AND REPRESENTATION

PHILOSOPHICAL STUDIES SERIES

VOLUME 99

Founded by Wilfrid S. Sellars and Keith Lehrer

The titles published in this series are listed at the end of this volume.

LANGUAGE, KNOWLEDGE, AND REPRESENTATION

Proceedings of the Sixth International Colloquium on Cognitive Science (ICCS-99)

Edited by

JESÚS M. LARRAZABAL

Institute for Logic, Cognition, Language and Information (ILCLI), The University of the Basque Country, Spain

and

LUIS A. PÉREZ MIRANDA

Institute for Logic, Cognition, Language and Information (ILCLI), The University of the Basque Country, Spain

KLUWER ACADEMIC PUBLISHERS

DORDRECHT / BOSTON / LONDON

A C.I.P. Catalogue record for this book is available from the Library of Congress.

ISBN 1-4020-2057-0 (HB)
ISBN 1-4020-2783-4 (e-book)

Published by Kluwer Academic Publishers,
P.O. Box 17, 3300 AA Dordrecht, The Netherlands.

Sold and distributed in North, Central and South America
by Kluwer Academic Publishers,
101 Philip Drive, Norwell, MA 02061, U.S.A.

In all other countries, sold and distributed
by Kluwer Academic Publishers,
P.O. Box 322, 3300 AH Dordrecht, The Netherlands.

Printed on acid-free paper

Printed in the Netherlands.

TABLE OF CONTENTS

INTRODUCTION

LANGUAGE, KNOWLEDGE, AND REPRESENTATION

The Sixth International Colloquium on Cognitive Science (ICCS-99), co-organized by the Institute for Logic, Cognition, Language, and Information (ILCLI) and the Department of Logic and Philosophy of Science, both from the University of the Basque Country, was held at Donostia - San Sebastián from May 12 to 19, 1999. In this occasion the main topics were the following ones:

1. Cognition: Epistemology and Metaphysics.

2. Cognition and Emotion.

3. Cognitive Approaches to Learning.

4. Foundations and Models of Discourse Pragmatics.

In a first rank forum about a hundred researchers exchanged their most recent contributions to Cognitive Science and its related areas. As in previous editions of this *Colloquium* all the sessions were held in a fruitful and friendly atmosphere. This volume contains a small but representative sample of the main papers presented at the conference. They all were invited papers except the one by Isabel Gómez Txurruka, a contributed paper that won the IBERDROLA Best Paper Award, given in ICCS-99 for the third time.

LANGUAGE, KNOWLEDGE, AND REPRESENTATION

One of the main goals in contemporary Cognitive Science is to increase our understanding of different cognitive processes, such as perception, learning, memory, reasoning, planning, decision-making, language acquisition, communication (literal and non-literal) and so forth. The diversity and use of different methods is nowadays manifest in all those disciplines that make up the wild spectrum of Cognitive Science. This last situation is in sharp contrast with the early days of the discipline when it was thought that the symbolic-representation was enough to do the job. Nowadays the current trend in

Cognitive Science is to work with several forms of representations and data. This change of outlook has led us to a kind of methodological *pluralism*. To be sure, psychological experiments, observations and descriptions of authentic cognitive processes in practical action or simulation of human behaviour are now common methods in the field.

Data learning is crucial for any cognitive or computational creature that tries to deal with practical issues in a changeable and dynamic world. However, most computational models of supervised learning rely only on labeled training examples, and ignore the possible role of unlabeled data. Tom M. Mitchell in his paper 'The Role of Unlabeled Data in Supervised Learning' presents an algorithm and experimental results demonstrating that unlabeled data can significantly improve learning accuracy in certain practical problems. He them identifies the abstract problem structure that enables the algorithm to successfully utilize these unlabeled data, and proves that unlabeled data will boost learning accuracy for problems in this class. This problem class includes many natural learning problems faced by humans, such as learning a semantic lexicon over noun phrases in natural language, and learning to recognize objects from multiple sensor inputs. The author argues that models of human and animal learning should consider more strongly the potential role of unlabeled data, and that many natural learning problems fit that class.

Even if the usual way of transmitting knowledge is through spoken or written natural language, there are many occasions where we need to enrich that language introducing more sophisticated symbols than ordinary ones. Formal and mathematical languages usually enable us to represent our abstractions and ideas, scientific concepts, and theories, etc., and manipulate them more accurately than by means of natural language itself. There are also images, graphs, tables, diagrams, etc. that are very useful for the transmission of ideas. These latter forms of representation provide a complementary and sometimes essential framework to ordinary languages. But, how do we represent, transmit or express our emotions to other people. There are words in our languages that don't represent knowledge, but emotions. Paul Ekman's paper 'Emotional and Conversational Nonverbal Signals' is devoted to revise and expand formulations which distinguished among a number of different types of body movements and facial expressions. In particular, he aims at distinguishing among different activities shown in facial and/or bodily movement, but which, as he points out, have quite different functions, origins and coding.

In addition to presenting explanatory models of cognitive abilities, it seems also convenient to examine and discuss those questions that give rise to fundamental philosophical issues for the cognitive sciences. Although situated at another level of analysis, several contributions to this volume show us that we cannot easily overlook the 'intrinsic' philosophical difficulties involved in the studies of human mind.

Martin Davies in his contribution 'Aunty's Argument and Armchair

Knowledge' presents what is known as Aunty's argument for the language of thought. According to this argument conceptualised thought requires the truth of the language of thought (LOT) hypothesis. The argument can be divided into two main steps: First, thinking involves the deployment of concepts and having concepts involves commitments to certain patterns of inference. Second, personal-level events of conscious judgement and thought are underpinned by occurrences of physical configurations belonging to kinds that figure in the science of information-processing psychology. M. Davies reviews Aunty's argument arriving to the conclusion that this argument takes us from the initial assumption of intentional realism to a version of the LOT hypothesis. Along the paper he also addresses a worry about *eliminativism*, and proposes a 'solution' to the problem of armchair knowledge.

Bill Brewer in his work 'Self-Knowledge and Externalism' discusses the possibility of combining a so-called 'externalist' theory of empirical content, on which the contents of a person's beliefs are determined in part by the nature of his extra-bodily environment, with a plausible account of self-knowledge, in particular, of a person's knowledge of the contents of his own beliefs.

How the psychological and the neurobiological investigations of cognitive phenomena should be related within a materialist approach to cognitive science is a much debated issue, and one of central importance for the more general question of the unity of the scientific study of cognition. Roy's paper, 'Cognitive Neuroscience and the Unity of the Study of Cognition', is an inquiry into the significance for this issue of the recent developments of cognitive neuroscience, broadly construed as that part of contemporary cognitive research investigating the neurobiological dimension of natural cognition. Such an inquiry is a standard case of epistemological analysis, in the sense of an attempt to clarify what scientific knowledge should be on the basis of an examination of what it is in a given body of scientific work. Roy's concern is mainly to lay the grounds for such an analysis. Accordingly, he first attempts to provide a rigorous specification of what it means and of the main tasks it involves, as well as to account for its necessity. Later he introduces a general framework for carrying out a systematic investigation of the different relations that should hold between the neurobiological and the psychological levels of inquiry into cognitive phenomena from a materialist viewpoint. On the basis of these preliminary developments, he finally sketchs out some elements of an epistemological analysis of contemporary developments of cognitive neuroscience, with a view to clarifying what they teach the materialist about the appropriate way of relating these two levels.

Consciousness is certainly one of the most perplexing and astonishing problems for the philosophy of mind, psychology and, probably, neuroscience. Ernest Sosa's paper 'Consciousness and Self-Knowledge' is devoted to what is known as *self-knowledge* problem. Sosa starts his analysis pointing out that Ryle's most famous and influential book, *The Concept of Mind*, is still today a very good reference for themes like the attack against dualism, and the various prongs of that attack, such as the distinction between know-how how and know-that. In Sosa's view, however, the arguments offered by Ryle against the doctrine of Privileged Access are somehow misleading. What is more, it is not clear that Ryle's definition properly identifies the most interesting form of that doctrine. Sosa introduces along the paper new conceptual elements that are very helpful at the understanding of the epistemology of self-knowledge.

William Lycan in his paper 'The Plurality of Consciousness' reveals some of the keys to look with new eyes at this old topic. On the one hand, he points out the necessity of distinguishing among the variety of topics and issues under the heading of «consciousness». In his opinion each of the problems related to that topic becomes more tractable once it has been carefully distinguished from the others. Some of these topics and issues are purely empirical, while others are more abstractly theoretical or conceptual and philosophical. On the other hand, he remarks that we cannot reduce the problem of «consciousness» to what is known as «the problem of phenomenal experience». In brief, there are two main morals that we could draw from his paper: First, no one should claim that problems of phenomenal experience have been solved by any purely cognitive or neuroscientific theory. Second, the theories cannot fairly be criticized for failing to illuminate problems of phenomenal experience.

Ronald de Sousa in his paper 'Twelve Varieties of Subjectivity: Dividing in Hopes of Conquest' presents different "senses" or "aspects" of subjectivity. He recognizes that some of them might be reduced or assimilated to others. But to establish conclusively that they are all distinct would involve 11! or 65 pairwise comparisons. His conclusion is that there are plausible, non-mystifying avenues of research open on each of the twelve forms of subjectivity described along the paper, and that in such piece-meal solutions lies the hope of solving the so-called problem of subjective consciousness.

In his paper 'On Collective Intention', Kepa Korta discusses several philosophical approaches to this notion (by Tuomela, Gilbert, Bratman, and Searle) that try to explain what is involved in collective agency and action. All the approaches considered by Korta are intended to reconcile the insight that there is such a thing as collective behaviour with the fact that, ontologically speaking, there is nothing over and above individuals, with their individual minds. According to Korta, some of these approaches attempt to reduce the notion of collective intention to a configuration of singular intentions plus mutual beliefs, but they fail, since they render circular (that is, non-reductive) definitions. Some other approaches seem to entail a non-mentalistic notion of (collective) intention, even if they do not clearly imply the existence of supra-individual minds. Korta's own conception favours a primitive notion of collective intention *à la* Searle. 'Collective intention' is a primitive notion that characterizes the mental states of the agents performing a collective action, distinct from a mere aggregate of individual actions. Korta gives some arguments for this approach and, at the same time, explores some properties of the notion of collective intention in relation to the notions of singular intention and mutual belief. At the end of his contribution, Korta makes some remarks towards a multimodal logic of collective intention.

In 'Focus in Discourse: Alternative Semantics versus a representational approach in SDRT', Isabel Gómez Txurruka discusses to what extent two different theories of focus, Rooth's Alternative Semantics and the approach developed in SDRT (Asher 1993), are appropriate to give an account of the interaction between focus and the linguistic environment. It is well known that the alternative framework postulates a semantic interpretation of focus, according to which the focus of a sentence is interpreted as a set of alternative propositions to the actual one. Completely different is the view taken in representationalist frameworks such as SDRT, where the focus/background structure introduces a simple partition of the representation contributed by the utterance. In this paper, the author identifies two possible problems the alternative framework has to deal with. The first is related to the fact that focus does not introduce an existential presupposition. She defends that, if this is so, the set of alternatives that constitutes the focus value cannot be related truth-conditionally to the previous discourse. The other difficulty has to do with the relation between the contribution of focus and the rest of the sentence in the focus value. The author suggests that this relation is not correct if the aim is to restrict the set of alternatives. In the second part of the paper, the focus-discourse interface is analyzed, paying particular attention to the interaction between focus and the discourse relation of Question-Answer Pair. The problems previously identified are exemplified and a very preliminary account of this interaction is offered in representational terms. Written with simplicity, this paper is well structured and offers several puzzles that will make the reader wonder about the relationship between focus and possible worlds and representationalist approaches to meaning.

In sum, this book collects the papers by some very well-known researchers in Cognitive Science, which represent leading edge work on *Knowledge, Language and Representation* from both a cognitive and a philosophical point of view. The diversity of their disciplinary origins and standpoints not only reflects the main topics and the range of different positions presented at this colloquium but also serves as proof of the richness, fruitfulness, and diversity of research in Cognitive Science today.

ACKNOWLEDGEMENTS

ICCS-99 was financially supported by the Secretary of State of Education and Culture (Ministerio de Educación y Cultura) of the Spanish Government, the Secretary of Education of the Basque Government, the Kutxa Foundation, and Iberdrola Electric Company. We would like to thank the members of the

Organizing and Program Committees of the Colloquium for their invaluable work before and during the Colloquium. Concerning the preparation of this volume, an anonymous referee and people from Kluwer, especially Keith Lehrer and Ingrid Krabbenbos must be thanked.

Jesús Mari Larrazabal and Luis A. Pérez Miranda

BILL BREWER

SELF-KNOWLEDGE AND EXTERNALISM

I want to discuss the possibility of combining a so-called 'externalist' theory of empirical content, on which the contents of a person's beliefs are determined in part by the nature of his extra-bodily environment, with a plausible account of self-knowledge, in particular, of a person's knowledge of the contents of his own beliefs. A difficulty for this combination is thought to be that it provides a wholly non-empirical source of knowledge about the mind-independent physical world which is intuitively intolerable.[1] The inference which is held to create this difficulty can be put like this.

(E1) I believe that *p*.

(E2) If *x* believes that *p*, then *x*'s environment is thus-and-so.

∴ (E3) My environment is thus-and-so.

An example of this might be the following.

(w1) I believe that water is wet.

(w2) If *x* believes that water is wet, then *x*'s environment contains (or did contain) water.

∴ (w3) My environment contains (or did contain) water.

Jesus M. Larrazabal and Luis A. Pérez Miranda (eds.), Language, Knowledge, and Representation: 1-18.

Note that the truth of the consequent of (w2) requires that *x*'s environment contains (or did contain) *water*, rather than any kind of 'twin-water', which is like water in all superficial respects but happens to have a different chemical composition - that is, it requires that *x*'s environment contains (or did contain) *this stuff* (H_2O) as opposed to *that stuff* (XYZ) (Putnam, 1975). This is precisely the force of the relevant form of content externalism. So the conclusion states that my environment contains *water*, as opposed to twin-water. This is a contingent matter of empirical fact, though. Hence the prospect of my knowing it without any kind of empirical investigation certainly raises a prima facie problem.

1

The proponent of this line of objection to combining an adequate account of self-knowledge with content externalism argues as follows. First, any adequate account of a person's knowledge of the contents of his own beliefs entails that his knowledge of instances of (E1) is *non-empirical*: neither its acquisition nor its status as knowledge necessarily involves any specific empirical investigation. Second, content externalism entails that instances of (E2) can be derived from non-empirical philosophical reflection, on the necessary conditions upon determinate empirical belief possession. Third, therefore, the truth of content externalism - in the presence of an adequate account of self-knowledge - enables a person knowledgeably to derive instances of (E3), on the basis of the argument above, without any empirical investigation whatsoever. Fourth, such non-empirical knowledge of empirical facts is intuitively intolerable. Therefore, fifth, content externalism is incompatible with any adequate account of a person's knowledge of the contents of his own beliefs. The reaction which is implicitly recommended by advocates of this argument, of course, is to reject content externalism. Like many others, though (e.g. Putnam, 1975; Burge, 1979, 1986; Evans, 1982; Pettit and McDowell, 1986; McGinn, 1989; Davies, 1991, 1992; Peacocke, 1993), I am very sympathetic to some form of externalism. Indeed, I think that *world-dependent* perceptual demonstrative contents are crucial, both to our very capacity to think about particular things in the world around us, and to our acquisition of knowledge about them (1999). So I need some alternative to the recommended reaction of rejecting such externalism.

Before considering what this should be, it is worth pointing out that some externalists have tried to exploit the intuitively intolerable result directly to their own advantage in presenting anti-sceptical transcendental arguments. "There is no such thing as a *reductio ad absurdum*", they say, "just the consequences of my view; and this one is very useful thank you". For example, Burge offers the following in this spirit.

(B1) I think that there are physical entities.

(B2) If x thinks that there are physical entities, then x bears causal-perceptual relations to physical entities.

∴ (B3) I bear causal-perceptual relations to physical entities.

I shall return to my assessment of this bullish attempt to turn the tables on critics of externalism when I have given my own account of how to respond to the apparent threat posed by arguments along these lines. Cutting to the chase: Burge's attempt fails.

What, then, are the externalist's options here? The one which I shall eventually propose and defend is to reject the very first move of the objection set out above. I deny that a person's knowledge of the contents of his own beliefs is entirely non-empirical in any way in which it really would follow that he could thereby acquire intolerably non-empirical knowledge of the contingent facts about his own environment. There is, in my view, an essential empirical component in self-knowledge. For determinate concept possession is, in a sense which I shall make precise, an *epistemic* skill. Before developing this claim, I want briefly to survey the alternatives.

First, it might be possible to argue that the content externalists not committed to the non-empirical knowability of *specific instances* of (E2). Although he is committed by definition to the conceptual necessity of the claim that concepts *of certain types* are externally individuated, in a way which entails the possibility of establishing that claim by non-empirical reflection, he may nevertheless deny that the question of which particular concepts instantiate these types can be settled without empirical investigation.[2] Second, it might also be possible to deny the third move above, according to which the truth of content externalism - in the presence of an adequate account of self-knowledge - enables a person *knowledgeably* to derive instances of (E3) without any empirical investigation. This is Martin Davies' response (1997), which I shall also consider in some detail shortly.

There are two further responses, which I shall simply mention and set aside without argument at this stage, as extremely unpromising. The first of these would be to insist, in extreme rationalist spirit, that there is no difficulty at all in the idea of wholly non-empirical knowledge of empirical matters of fact. The final status of the contrary intuition, against this possibility of natural science by non-empirical reflection, is bound up, of course, with the fate of anti-sceptical transcendental arguments like Burge's, above, to which I shall return. The second unpromising response would be to claim that a person's self-ascriptions of beliefs are not, contrary to appearances, genuinely truth-evaluable *statements*, but rather non-truth-evaluable avowals of some kind (Wittgenstein, 1958, pp. 190-192; 1980, §§ 470-504; Malcolm, 1991). Thus, they are incapable of constituting the premises of an argument, as they are supposed to do in (E1), the first premise of the argument which is in turn supposed to cause trouble for content externalism. Again, this move seems desperate at this point. Whatever their logical and epistemological peculiarities, self-ascriptions of beliefs surely are statements about a person, the person making them, to the effect that he or she is in a certain condition, namely that of believing that *p*, say.

2

Let me return, then, to the two more plausible alternatives identified above to my own preferred strategy for establishing the compatibility of self-knowledge and content externalism in the face of the present line of objection. The first of these denies that specific instances of (E2) are knowable non-empirically, whilst granting that content externalism itself is, by arguing that it is a matter of empirical investigation to *which* actual concepts this externalism applies. For the standard externalist examples are natural kind concepts. Yet knowledge that any given concept is a genuine natural kind concept requires knowledgeably ruling out the following two ways in which it might fail to be so, which is surely an empirical matter.

First, what appears to be reference to a natural kind may fail to be so because the characteristic theoretical role associated with the putative kind is not in fact played by anything at all. 'Phlogiston' provides a familiar actual case of this type. There just is no kind of stuff which is released into the surrounding medium quickly by burning and slowly by rusting, and so on. These processes involve instead the absorption of oxygen from the atmosphere. So there is no such thing as phlogiston. Apparent reference to a natural kind by the use of that term is merely illusory, and therefore certainly not susceptible to externalist treatment. Call this *the possibility of emptiness* with respect to a given putative natural kind term.

Second, the range of items to which a term is correctly applied may turn out to be quite heterogeneous at the relevant level of theoretical description involved in the discrimination and unification of natural kinds. 'Jade' provides a familiar actual example of this phenomenon; for it is correctly used of two quite different substances: jadeite, which is a silicate of sodium and aluminium; and nephrite, which is a silicate of lime and magnesia. Cases of this kind could clearly turn out to be far more wildly heterogeneous, sufficiently so to make quite implausible any attempt to think in terms of a single unified natural kind with a number of sub-varieties, as it may be correct to think that some water is H_2O and some is D_2O, although both are sub-varieties of the single natural kind water, and not of the natural kind twin-water (XYZ). In the more extreme cases, again, what appears to be a natural kind term turns out, on empirical investigation, not to be so. Call this *the possibility of heterogeneity*.

Thus, it may be claimed, knowledge of any particular instance of (E2) depends upon knowledge that the relevant component concept of the content that *p* is a *genuine* natural kind concept, which in turn depends upon the *empirical* knowledge required to rule out the possibilities of emptiness and heterogeneity with respect to the term expressing that concept. So the argument from (E1) and (E2) to (E3) does not, after all, threaten to provide an untenable, because wholly non-empirical, source of any specific piece of empirical knowledge.

I think that the content externalist should not be overly impressed by this apparent resolution of the tension between his position and the possibility of a plausible account of self-knowledge. The way in which parallel issues arise in connection with perceptual demonstrative reference to particular objects helps to bring out why. Consider, for example, the following instance of the argument (E1),(E2) |- (E3).

(d1) I believe that that ball will go into the pocket.

(d2) If *x* believes that that ball will go into the pocket, then *x*'s environment contains that ball.

∴ (d3) My environment contains that ball.

Of course, there is no threat that such an argument should constitute an untenably non-empirical source of empirical knowledge - that a particular ball exists in his environment - if the subject's attempted demonstrative reference fails. And there are at least the following two ways in which this might come about. First, there maybe no ball at all, or anything else for that matter, at the place in his environment where the subject takes there to be one. That is to say, he maybe subject to some kind of hallucination. Second, his thinking may fail to be responsive to the behaviour of a single ball, as he fails, for example, to keep track of the movement of a single red ball after a very strong break in a game of snooker. In the absence of any such defeating abnormality, though, the argument is perfectly sound; and neither his successful demonstrative reference to the particular ball in question, nor his knowledge of the particular instance of (E2) - that is, his knowledge of (d2) above - depends upon his having carried out any prior, or independent, empirical investigation to rule out all the possible sources of hallucination, unnoticed substitution of one ball by another, failure of attentional tracking, and so on.

Similarly, I contend, in the case of reference to natural kinds. According to the most natural development of the externalist position, understanding of natural kind terms is, in the most basic cases, acquired by some kind of demonstrative identification of the kind in question, as *'that stuff'* - 'that liquid', say - on the basis of perception of its instances and some grasp of their distinguishing features and characteristic behaviour. The possibilities of emptiness and heterogeneity with respect to putative natural kind terms arise as follows. In the first case, the *appearance* of an underlying kind is entirely hallucinatory. In the second case, the attempted demonstrative identification fails sufficiently to keep track of any *single* kind. Either way, the relevant instance of the argument (E1), (E2) |- (E3) is quite harmless to the content externalist: its second premise is simply false. Nevertheless, in the normal case, when the demonstrative identification crucial to the subject's understanding of the term in question makes successful reference to a genuine natural kind, neither this, nor his knowledge of the corresponding specific instance of (E2), depends upon his having carried out any prior, or independent, empirical investigation to rule out all the possible sources of error due to emptiness or heterogeneity. Thus, the relevant instance of (E1), (E2) |- (E3) is back in contention as a threat to content externalism, appearing, as it does, to provide an untenably non-empirical source of empirical knowledge. In any case, the present attempt to avoid this difficulty for combining content externalism with an adequate account of self-knowledge is unsuccessful.

The second prima facie promising alternative to my own account of how the two are to be combined is to appeal to Davies' (1997) claim that, regardless of the validity of the argument from (E1) and (E2) to (E3), of a person's knowledge of instances of its two premises, and, indeed, of his knowledge of the validity of the relevant form of argument itself, knowledge of instances of its conclusion still *cannot* be acquired by these means. For this would be in contravention of the following Limitation Principle.

(LP) Epistemic warrant cannot be transferred from A to B, even given an a priori known entailment from A to B, if the truth of proposition B is a precondition of the knower being able to believe the proposition A.

The only motivation which he offers for this principle is that it is supposed to save him from precisely the present difficulty with reconciling self-knowledge and externalism; and it looks a little bit like the following principle which he reads into Wittgenstein's *On Certainty* (1975).

(WP) Epistemic warrant cannot be transferred from A to B, even given an a priori known entailment from A to B, if the truth of proposition B is a precondition of the warrant for A counting as a warrant.

I think that neither is what Wittgenstein has in mind. More importantly, neither is remotely plausible in my view: (LP) even less so than (WP). A whole class of serious counterexamples to (LP) come immediately to mind. Consider the following inference.

(M1) r_1 and r_2 are real numbers

(M2) The product of any two real numbers is a real number

∴ (M3) r_1r_2 is a real number

Its conclusion is a necessary truth, which is therefore, presumably, a precondition of anything possible. Hence it is certainly a precondition of a person's believing the premises (M1) and (M2). So it follows from (LP) that epistemic warrant cannot be transferred from (M1) and (M2) to (M3). Generalizing the case, (LP) has the consequence that inferential knowledge of any necessary truth is impossible, which certainly places logic, mathematics, and, on many conceptions, philosophy, in a very poor position epistemologically speaking. Something must have gone badly wrong with (LP). So Davies' escape route for the externalists closed.

3

The key to my own strategy for reconciling self-knowledge with content externalism lies in the idea that true content externalist requirements are a consequence of the following version of Russell's Principle of Acquaintance (1917, p. 159), which I shall call (A).

> (A) A person's capacity to make determinate reference to certain objects and kinds in belief depends upon his having *demonstratively-based knowledge* about them.

This is what makes it the case that his possession of such beliefs depends upon his being in an environment actually containing the objects and kinds in question. What does this acquaintance condition amount to, though? Well, a person has *demonstratively-based knowledge* about a given object or kind just if, either, he has knowledge expressible using demonstrative reference to that object or kind, or, he had such knowledge, some of which is retained, perhaps linguistically categorized and no longer *demonstratively* expressible, in memory.

Establishing this principle (A) obviously requires extended argument, of which I only offer a sketch here.[3] The crucial claim is that externalist relations are necessarily *reason-giving* relations, constituting a source of demonstratively expressible knowledge; where by this I mean reason-giving in the epistemological internalist's sense: reason-giving *from the subject's point of view*, rather than merely from the perspective of some external theorist. Let me explain what this means and why I endorse it.

The externalist holds that certain empirical beliefs have their contents determined in part by the subject's relations with particular things in his environment, the causal relations, for example, in which he stands to such things - to that object, say, or (instances of) that natural kind - when he is perceiving them, being informed about them by others, remembering them, and so on. Call these his *causal-perceptual relations* with the external worldly things in question. Now, suppose that these content-determining causal-perceptual relations are *not* reason-giving relations; and consider a person, *S*, who believes that *p*, where this is supposed to be an empirical belief with externalistically determined content. Since his causal-perceptual relations with the things around him play an *essential* role in the determination of the contents of his empirical beliefs, on the externalist's account, there is a range of alternative such beliefs - beliefs which he might have had instead - whose difference in content with his actual belief that *p* would have been due *entirely* to his standing in the relevant content-determining causal-perceptual relations with different mind-independent things. Suppose that the belief that *q* is one of these.

So, the situation is this. *S* actually believes that *p*, because his actual environment determines this, as opposed to *q*, as the empirical content of his belief, through its causal-perceptual impact upon him. He does not believe the distinct content that *q*. Had his environment been appropriately different, though, his position would have been precisely the reverse: he would have believed that *q*, and not believed that *p*. Yet, by hypothesis, the relevant content-determining causal-perceptual relations in which he stands to the actual things in the world around him are *not* reason-giving relations. So his standing in them is *quite neutral*, from his point of view, on which such things it is to which he is so related. For if they were not neutral in this way, then they would be reason-giving in the relevant sense. Thus, his standing in these relations is quite neutral, from his point of view, on whether it turns out that he is believing that *p* or that *q*. Yet they are the only difference between the actual case, in which he believes that *p*, and the counterfactual case, in which he believes that *q* - for this is how *q* was introduced in relation to *p*. Thus, there could not possibly be, on this account, any difference at all, from the subject's point of view, between his believing that *p* and his believing that *q*.

Which of the two beliefs he actually has is due entirely to which environmental entities it happens to be in which he stands in the relevant causal-perceptual relations; yet these relations are quite neutral, from his point of view, on which entities these turn out to be. Any supposed difference between believing that *p* and believing that *q* is therefore *absolutely nothing to him*. So he does not really *understand* them *as alternatives*. Believing that *p* and believing that *q* are identical for him. Hence the supposedly content-determining role of *S*'s environment is empty. For there is nothing more, or less, to the content of a belief than the way *the subject* takes the world to be. Thus, if the proposed causal-perceptual relations in which a person stands to certain mind-independent things are not reason-giving relations, then they contribute *nothing* to the determination of specific truth-conditions for his empirical beliefs. In other words, the content-determining relations between a person and certain things in the world around him which are posited by the content externalist must be reason-giving relations.

The form of this argument is that of what Peacocke calls 'the *switching* tactic' (1988, pp. 475 ff).[4] A more familiar historical paradigm is provided by Strawson's (1959, ch. 3; 1966, pt. III, sect. II, esp. pp. 168 ff; and 1974) reading of Kant's (1929, A341/B399 ff) objection to Descartes' (1986, pp. 107 ff) substance dualism. According to this Kantian argument, substance dualism entails the coherence of a distinction between qualitative and numerical identity for immaterial minds; yet the dualist's own conception of such things, as exhaustively characterized by what is infallibly given to their own subjective point of view, denies her the resources to give any genuine content to the idea of two qualitatively identical but numerically distinct minds. So, the substance dualist depends upon a distinction - between qualitative and numerical identity for immaterial minds - which she is, by her own lights, incapable of making. Thus, the position is internally inconsistent.

Similarly, here, we are to consider a theorist who insists that the non-reason-giving causal-perceptual relations in which a person's stands to certain objects or kinds in his worldly environment are essential to the determination of specific contents for his empirical beliefs. Such a theorist is therefore committed to the existence of pairs of beliefs with genuinely distinct contents, the distinction between which is entirely due to their subjects' standing in the relevant causal-perceptual relations with different possible objects or kinds. Given her own conception of the nature of these content-determining relations, though, as *non-reason-giving* relations, this entails an overall conception of belief content which countenances the following situation. A person actually believes that *p* (and does not believe that *q*). He might instead have believed that *q* (and not believed that *p*). Yet everything involved in determining the actual case as one of his believing that *p*, not that *q*, is entirely neutral, from the subject's point of view, between this situation and the reverse, in which he believes that *q*, not that p. In other words, non-reason-giving content-determining relations of this kind induce distinctions in belief content which are more discriminating than anything which could possibly enter into the subject's own understanding of his beliefs. The content of a belief is precisely, no more and no less than, the way the subject understands things to be. Yet the theorist under consideration here is committed incoherently to making distinctions in belief content which she is incapable by her own lights of making in these terms.

There is a line of thought in the recent externalist literature which is supposed to address just this kind of worry (see Burge, 1996; and Peacocke 1996). It suggests that a person's own grasp of his externally determined belief contents consists in his ability authoritatively to self-ascribe precisely those contents, whichever they turn out to be. Suppose that *S*'s actual belief that p and counterfactual belief that *q* would each be expressed by the sentence '*K* is *F*'. Their difference in content resides in the fact that the kind term '*K*' has a different semantic value in each case, where the semantic value of such a term is the natural kind in the world whose association with that term determines its contribution to the truth-conditions of propositions expressed using it. In the actual case, '*K*' has semantic value k_1; in the counterfactual case, its semantic value is k_2. Now, the current line of thought is that, since the same term is used in the subject's self-ascription as is used in expressing the first order belief which he thereby self-ascribes, this self-ascription is bound to be correct in either case. So his capacity authoritatively to make it constitutes his *understanding* of his actual belief that p, in a way which distinguishes this from the counterfactual belief that *q*. In self-ascribing as follows: 'I believe that *K* is *F*', in the actual case, in which the semantic value of '*K*' is k_1, he correctly self-ascribes his belief that *p* and not the belief that *q*, although he equally correctly self-ascribes the distinct belief that *q*, in the counterfactual case in which '*K*' has semantic value k_2. So the difference between the cases in which he believes that *p* and believes that *q* does, after all, show up from his point of view. He distinguishes them by being able correctly to make different self-ascriptions in each.

I find this line of thought quite unconvincing. Unless he already knows which kind *K* actually is - in some sense, that the semantic value of '*K*' is k_1 as opposed to k_2 - he cannot hope to inform himself of this just by reusing the word in an attempt to tell himself which it is. This so-called 'conceptual redeployment' routine is effective only if the subject can distinguish the two self-ascriptions which he is supposed to make, in the actual and counterfactual cases. Yet this presupposes his *prior* grasp of the distinction between believing that *p* and believing that *q*. So it cannot be brought in after the event in an attempt to account for this understanding of the distinct first order beliefs. The subject's insensitivity to the semantic value of' *K*' - that this is k_1 as opposed to k_2 - at the first order level is simply recycled in his second order self-ascription, which he therefore equally fails to *understand*. Although the sentence he uses to self-ascribe his belief will, on this view, be true in both the actual and counterfactual cases, the subject himself is left hopelessly ignorant of which truth this turns out to be in either case.

So, the Switching Argument stands. If a person's causal-perceptual relations with mind-independent objects and kinds are to contribute essentially to the determination of the empirical contents of his beliefs, then his standing in these relations must provide him with reasons for such beliefs: beliefs, that is to say, about just *that object*, or *that kind*. These content-determining causal-perceptual relations therefore constitute a source of demonstratively expressible knowledge about such objects and kinds: the very objects and kinds which are the semantic values of the relevant externalist concepts. Hence the source of externalist conditions upon determinate empirical belief possession is indeed my version of Russell's Principle of Acquaintance, (A) above. The requirement upon possession of empirical beliefs with certain contents that the subject should be in an environment actually containing certain objects or kinds is derived from the fact that his possession of such beliefs depends upon his actually standing in certain basic reason-giving relations with such things, which serve as a source of demonstratively-based knowledge about them.

I think that considerations from a slightly different area lend additional support to (A). Recall, first, Evans' comments about a person's use of the proper name 'Louis' in the course of his discussion of the causal theory of names (1985a).

> A group of people are having a conversation in a pub, about a certain Louis of whom *S* has never heard before. *S* becomes interested and asks: 'What did Louis do then?' There seems to be no question but that *S* denotes a particular man and asks about him. Or on some subsequent occasion *S* may use the name to offer a new thought to one of the participants: 'Louis was quite right to do that.' Again he clearly denotes whoever was the subject of the conversation in the pub. This is difficult to reconcile with the Description Theory [on which there is supposed to be associated with each name as used by a group of speakers who believe and intend that they are using the name with the same denotation a description or set of descriptions cullable from their beliefs which an item has to satisfy to be the bearer of that name] since the scraps of information which he picked up during the conversation might involve some distortion and fit someone else much better. Of course he has the description 'the man they were talking about' but the theory has no explanation for the impossibility of its being outweighed.
>
> The Causal Theory [on which it is sufficient for someone to denote *x* on a particular occasion with a name, that this use of the name on that occasion be a causal consequence of his exposure to other speakers using the expression to denote x] can secure the right answer in such a case but I think deeper reflection will reveal that it too involves a refusal to recognize the [Wittgensteinian] insight about contextual definition [that for an item to be the object of some psychological state of yours may be simply for you to be placed in a context which relates you to that thing] .... For the theory has the following consequence: that at any future time, no matter how remote or forgotten the conversation, no matter how alien the subject matter and confused the speaker, *S* will denote one particular Frenchman - Perhaps Louis XIII - so long as there is a causal connection between his use at that time and the long distant conversation. (pp. 6-7)

Evans has two important points here: first, that possession of a uniquely identifying definite description is unnecessary for successful singular reference; second, that the mere existence of a causal chain of 'reference-preserving' links back to the object in question, as these are conceived, for example, by Kripke (1980) and other proponents of the 'causal theory of reference', is insufficient. Our intuitions about the 'Louis' case surely confirm both of these points. My hypothesis is that these intuitions are organized and controlled precisely by the existence of *epistemic* constraints upon successful reference in thought. What makes *S*'s context in the pub conversation sufficient for him to denote Louis XIII in this way, is that he is there, at that time and in the context of that conversation, in possession of demonstratively-based *knowledge* about that man. His grasp of what is being said by those around him, and his understanding engagement in the discussion generally, provide him with knowledge expressible using demonstrative reference to the person in question: 'that guy (Louis) had a hard time', say.[5] Equally, I contend, what denies his later uses of any such significance, in the circumstances which Evans describes, is that he then no longer retains anything of this *knowledge* in memory.

The point here can be generalized. One source of illumination for the relation of reference - both to particular objects and to natural kinds - which holds between certain referring expressions, as they are used in a given linguistic community, and the things to which they refer, lies in reflecting upon the practice of the radical interpreter in formulating a truth-theory for the language in which such expressions occur (see Davidson, 1984, esp. essays 3, 9, 10, 11, 14, 15 and 16; Evans and McDowell, 1976, intro; McDowell, 1977, 1978). Thinking in this way enables one to give a perfectly adequate account of the relations between thinkers and things which are required if the former are to refer to the latter in thought and talk, whilst resisting any need reductively to formulate this account in terms of certain specific types of causal relations, or relations of any other kind, conceivable quite independently of their role in the intelligible engagement of a rational agent with the world around her.

The relations, then, between a person, in her use of a referring expression, and the thing to which she refers in using it, in virtue of which the expression does indeed refer to that thing, are precisely those relations which prompt an ideal interpreter, in his attempts to make best overall sense of what she says and does, to regard her as talking about that thing in using the expression in question; and there may be no other way to identify or characterize these relations. My proposal is that this process of making sense of what people are thinking and talking about is constrained precisely by considerations of what they have, or could have, *knowledge* about, most importantly, what it is about which they are provided with demonstratively expressible knowledge by the way in which their attention is focused upon the world in perception, testimony, memory and so on: given, that is, *the world as it presents itself to their point of view*. Thus, I claim that the relevant process of interpretation is governed by the question which things in the world around them the subjects to be interpreted could have demonstratively-based *knowledge* about, given the way they are built, the relations in which they stand to such things in using the linguistic expressions which they use in the ways in which they do, and the other ways in which they operate in the world around them.

In other words, there are, amongst the factors determining the interpreter's assignment of a particular object, or natural kind, as the reference of a given referring expression in use in a certain linguistic community, significant *epistemic* constraints of precisely the kind which my reading of Russell's Principle of Acquaintance, (A) above, demands. The sort of engagement between the language-users and the particular object or kind in question which is required if this assignment is really to make best sense of what they say and do is precisely that involved in their acquiring *knowledge* about just *that object*, or *that kind*, and retaining this to some extent in memory, in the circumstances of their use of the expression. This is my principle (A): reference to certain objects and kinds requires demonstratively-based knowledge about them. Put another way, the claim is that a person succeeds in singling out a determinate particular or kind in thought, in each of the wide variety of modes of reference by which this is possible, only in virtue of his standing in certain relations with that object or kind which provide him with demonstratively-based knowledge about it. These *epistemic* constraints upon reference are what, as McDowell puts it, "anchor" the correct semantic theory for a given language to the actual facts of its use (1977, pp. 183-184).

What remains to be shown is exactly how this basis for content externalism, in the Russellian thesis (A), undermines its purported incompatibility with any adequate account of self-knowledge.

Suppose that a person's belief that *p* comprises an externalist concept *C*. The challenging inference would then be this.

(e1) I believe that *p*.

(e2) If *x* believes that *p*, then *x*'s environment contains (or did contain) *C*.

∴ (e3) My environment contains (or did contain) *C*.

On my view, as I say, the externalist requirement upon possession of the concept *C* derives from the fact that its semantic value is necessarily a natural kind, say, about which any person who has the concept has demonstratively-based knowledge, where what this amounts to is either that he has knowledge expressible using demonstrative reference to that kind - e.g. 'that is water in the glass over there', or 'that water looks refreshing' - or that he had such knowledge, some of which is retained, perhaps linguistically categorized and no longer demonstratively expressible, in memory - e.g. cool water is refreshing to drink', or 'the water in Hinksey pool was very cold'. Now, if the inference set out above is to be an unwarrantedly non-empirical source of knowledge, then its premises must at least be true. The truth of (e1) depends upon the subject's grasp of the content '*p*', though, which in turn depends upon his possession of the concept *C*. From (A), he therefore has demonstratively-based knowledge about *C*, along the lines suggested above. Hence he is already in a position to arrive at the knowledge that there is (or was) *C* in his environment if only he turns his mind to the matter. For example, if he already knows that that is water in the glass over there, or that the water in Hinksey pool was very cold, his knowledge that there is water in his environment cannot possibly be any mystery. Therefore this argument cannot possibly constitute a problematic non-empirical *source* of new empirical knowledge: if its premises are simply true, then the subject already has the wherewithal to arrive at knowledge of its conclusion.

Put slightly differently, the key claim is that the opponent of externalism wrongly neglects the empirical-epistemic constraints upon concept possession which essentially enter into a person's knowledge of (el) through their application simply to its truth. This already presupposes his standing in an *epistemic* relation - Russell's *acquaintance*, as it were - with samples of *C*. It is the first move in the objector's reasoning at the outset, then, that any adequate account of a person's knowledge of the contents of his own beliefs entails that this is *wholly non-empirical*, which is to be rejected. For this self-knowledge requires his grasp of the contents of the beliefs in question, his possession of whose component concepts in turn depends upon his empirical-epistemic relations with their semantic values. Thus, content externalism of this kind is perfectly compatible with an adequate account of a person's knowledge of the contents of his own beliefs.

To repeat, determinate concept possession is an *epistemic* skill. It is a matter of a person's being in relations with the relevant worldly semantic values which provide him with demonstratively-based knowledge about such things. This is the source of the externalist requirements upon concept possession. That is to say, the world-involving causal-perceptual relations between a person and certain things in the world around him which are essential to his possession of concepts with those things as their determinate semantic values are precisely the *reason-giving* relations which undermine the purported difficulty with which I began for combining content externalism with a plausible account of self-knowledge.

Finally, what are the consequences of all of this for Burge's, and others', anti-sceptical strategy? It faces the following dilemma. Either the causal-perceptual relations governing the content externalism of (B2) are reason-given relations or they are not. If they are, then the transcendental argument is unnecessary. For, as above, the truth of the first premise depends upon the subject's actually standing in such relations with particular physical entities in the world around her, which already provide her with reasons for certain beliefs about them. She has no need for the detour via Burge's argument. This is precisely why the arguments I have been considering pose no threat to externalism. For they do not constitute a non-empirical *source* of new empirical knowledge: subjects who can run them already effectively have it. If, on the other hand, the relevant causal-perceptual relations are not reason-giving, then the transcendental argument is completely ineffective. For, as I remarked earlier, the conceptual redeployment routine to which Burge appeals in explaining the subject's knowledge of (B1) delivers only her knowledge that the sentence "I am thinking that there are physical entities" is true, in a sense which is neutral between the cases, (a), in which 'physical entities', on her lips, refers to genuine physical entities, because they are what causally explain her use of the relevant expressions, and (b), in which 'physical entities', on her lips, refers instead to the deceptive intentions of the sceptic's malicious demon, say, or the brain stimulations of the mad scientist. Although she may therefore derive, by Burge's argument, the conclusion that she stands in causal-perceptual relations with those things, whichever they are, to which her use of expressions for physical entities is her normal response, she hasn't the slightest idea what kinds of things these are: in particular, whether they are genuine physical entities or rather some kind of sceptical surrogates for these. Either way, then, the anti-sceptical strategy fails.[6]

Bill Brewer
St. Catherine's College (Oxford)
U.K.

NOTES

[1]The particular formulation of the issue which I address here is due to Martin Davies (1997). Indeed, my thoughts in this area were initially stimulated by the opportunity to give a reply to an earlier version of his paper at a meeting of the European Society for Philosophy and Psychology in Barcelona during the summer of 1996. Paul Boghossian also presses the issue as a challenge to content externalism of the kind defined in the text below (1989, 1997).
[2]Christopher Peacocke urged me to take account of this possibility in correspondence.
[3]The argument is developed in detail, given extended illustration, and defended against putative counterexamples in ch. 3 of my forthcoming book (1999).
[4]So-called, I presume, because the tactic is to object to a theory on the grounds that it is in principle incapable of giving any significance to switches which are by that theory's own lights crucial.
[5]I realize that I have said nothing about how exactly testimony might provide such direct demonstrative *knowledge* about the objects of discussion in certain

circumstances. I believe that it does; and I would expect the correct account of this possibility to emerge from an investigation into the way in which what might be called *testimonial demonstratives* - as, for example, when *S* says '*that man* was a villain' in the context of Evans' case of the pub conversation about Louis XIII - succeed in referring to particular persisting mind-independent objects. This would be very much in accord with the way in which I argue elsewhere that an account can be derived of the way in which perceptual experiences provide non-inferential *reasons* for empirical beliefs from reflection upon perceptual demonstrative *reference* to mind-independent things (1999, esp. ch. 6). See Evans (1982, chs. 5 &9); Fricker (1987); Coady (1992); and McDowell (1994) for important work on the epistemology of testimony.

[6]This paper inherits many acknowledgments from my forthcoming book and has acquired its own too. I should like to thank the following for their very helpful comments on earlier versions of this material. John Campbell, David Charles, Bill Child, Martin Davies, Naomi Eilan, Jennifer Hornsby, Brian McLaughlin, Mike Martin, Christopher Peacocke, Johannes Roessler, Helen Steward, Tom Stoneham, Rowland Stout, Ralph Walker and Tim Williamson.

REFERENCES

Boghossian, P. 1989. 'Content and Self-Knowledge'. *Philosophical Topics*, **17**, 5-26.

— 1997. 'What the Externalist Can Know A Priori'. *Proceedings of the Aristotelian Society*, **97**,161-175.

Brewer, B. 1999. *Perception and Reason*. Oxford: Oxford University Press.

Burge, T. 1979. 'Individualism and the Mental'. In French et al (eds.), *Midwest Studies in Philosophy*, **4**,73-121.

— 1986. 'Individualism and Psychology'. *Philosophical Review*, **95**, 3-45.

— 1996. 'Our Entitlement to Self-Knowledge'. *Proceedings of the Aristotelian Society*, **96**, 91-116.

Coady, C. A. J. 1992. *Testimony*. Oxford: Oxford University Press.

Davidson, D. 1984. *Inquiries into Truth and Interpretation*. Oxford: Oxford University Press.

Davies, M. 1991. 'Individualism and Perceptual Content'. *Mind*, **100**, 461-484.

— 1992. 'Perceptual Content and Local Supervenience'. *Proceedings of the Aristotelian Society,* **92**, 21-45.

— 1997. 'Externalism, Architecturalism and Epistemic Warrant'. In C. MacDonald, B. Smith and C. Wright (eds.), *Knowing Our Own Minds*. Oxford: Oxford University Press.

Descartes, R. 1986. *Meditations on First Philosophy*, tr. J. Cottingham. Cambridge: Cambridge University Press.

Evans, G. 1982. *The Varieties of Reference*. Oxford: Oxford University Press.

— 1985a. 'The Causal Theory of Names'. In his *Collected Papers*. Oxford: Oxford University Press.

— 1985b. 'Understanding Demonstratives'. In his *Collected Papers*. Oxford: Oxford University Press.

Evans, G. and McDowell, J. (eds.). *Truth and Meaning*. Oxford: Oxford University Press.

Frege, G. 1993. 'On Sense and Reference'. In A. Moore (ed.), *Meaning and Reference*. Oxford: Oxford University Press.

Fricker. E. 1987. 'The Epistemology of Testimony'. *Proceedings of the Aristotelian Society, Supplementary Volume*, **61**, 57-83.

Kant, I. 1929. *Critique of Pure Reason*, tr. N. Kemp Smith, London: Macmillan.

Kripke, S. 1980. *Naming and Necessity*. Oxford: Blackwell.

Malcolm, N. 1991. '"I Believe that p"'. In E. LePore and R. Van Gulick (eds.), *John Searle and his Critics*. Oxford: Blackwell.

McDowell, J. 1977. 'On the Sense and Reference of a Proper Name'. *Mind*, **86**, 159-185.

— 1978. 'Physicalism and Primitive Denotation: Field on Tarski'. *Erkenntnis*, **13**, 131-152.

— 1994. 'Knowledge by Hearsay'. In B. K.Matilal and A Chakrabarti (eds.), *Knowing from Words*. Amsterdam: Kluwer.

McGinn, C. 1989. *Mental Content*. Oxford: Blackwell.

Peacocke, C. 1988. 'The Limits of Intelligibility: A Post-Verificationist Proposal'. *Philosophical Review*, **97**, 463-496.

— 1993. 'Externalist Explanation'. *Proceedings of the Aristotelian Society*, **93**, 203-230.

— 1996. 'Entitlement, Self-Knowledge and Conceptual Redeployment'. *Proceedings of the Aristotelian Society*, **96**, 117-158.

Pettit, P. and McDowell, J. (eds.). 1986. *Subject, Thought and Context*. Oxford: Oxford University Press.

Putnam, H. 1975. 'The Meaning of "Meaning"'. In his *Mind, Language and Reality*. Cambridge: Cambridge University Press.

Russell, B. 1917. 'Knowledge by Acquaintance and Knowledge by Description'. In his *Mysticism and Logic*. London: Allen and Unwin.

Strawson, P. F. 1959. *Individuals*. London: Methuen.

— 1966. *The Bounds of Sense*. London: Methuen.

— 1974. 'Self, Mind and Body'. In his *Freedom and Resentment and Other Essays*. London: Methuen.

Wittgenstein. 1958. *Philosophical Investigations*, tr. G. E. M. Anscombe. Oxford: Blackwell.

— 1975. *On Certainty*, ed. G. E. M. Anscombe and G. H. Von Wright, tr. D. Paul and G. E. M. Anscombe. Oxford: Blackwell.

— 1980. *Remarks on the Philosophy of Psychology*, vol. I, ed. G. E. M. Anscombe and G. H. von Wright, tr. G. E. M. Anscombe.

MARTIN DAVIES

AUNTY'S ARGUMENT AND ARMCHAIR KNOWLEDGE

In my contribution to the Proceedings of the First International Colloquium on Cognitive Science, held in Donostia (San Sebastián) in 1989, I advanced what I called 'Aunty's own argument for the language of thought'.[1] The Aunty in question is Jerry Fodor's. He represents her as a conservative figure who is more likely to favour connectionism than to accept that there are good reasons to adopt the language of thought hypothesis.[2] As I envisaged her, she has some sympathy for the views of the later Wittgenstein but is fundamentally a neo-Fregean. I claimed that the neo-Fregean framework offers Aunty the resources to construct her own argument for the language of thought hypothesis, an argument that is relatively non-empirical in character.[3]

Aunty's own argument is an argument for the claim that conceptualised thought requires the truth of the language of thought (LOT) hypothesis. It proceeds in two main steps. The first step makes use of neo-Fregean resources: thinking involves the deployment of concepts and having concepts involves commitments to certain patterns of inference. In particular, it is assumed that conceptualised thought involves performing certain inferences *in virtue of their form*. The second step makes use of a quite general connection between tacit knowledge of rules and syntactically structured representations. A background assumption for the whole argument is that personal-level events of conscious judgement and thought are underpinned by occurrences of physical configurations belonging to kinds that figure in the science of information-processing psychology. These physical configurations can be assigned the contents of the thoughts that they underpin. They are 'proposition-sized' bearers of causal powers. This assumption is what Fodor (1985, 1987) calls *intentional realism* and it is close to the assumption of *propositional modularity*.[4] In my view, we are committed to this assumption by some of our everyday practices of mental talk and explanation, but I shall not spell out the nature of this commitment here.

Jesus M. Larrazabal and Luis A. Pérez Miranda (eds.), Language, Knowledge, and Representation: 19-37.

In this paper, I briefly review Aunty's argument and then respond to an objection, address a worry, and propose a solution to a serious problem.[5]

1. A BRIEF REVIEW OF AUNTY'S ARGUMENT

In order to think particular types of thought (such as the thought that 7+5=12) a thinker must possess particular concepts (here, the concepts of addition and of equality and concepts of the numbers 5, 7 and 12). Furthermore, possessing particular concepts involves a thinker in commitments to particular forms of inference. This is not to say that there are forms of inference that the thinker will invariably accept when they are presented or perform when they are appropriate. Possessing a concept is a matter of having certain cognitive abilities, but there are many factors that may prevent an ability from being exercised. In Chomsky's (1965) terminology, possessing a concept is a matter of competence rather than of performance.

Commitment to a particular form of inference is not just commitment to each of a number of inferences that happen to instantiate that form. Rather, the commitment is to accept or perform those inferences in virtue of their form. The form of the inferences should figure, somehow, in the causal explanation of the thinker's performing those inferences.[6] But it is not obvious what this requirement comes to. We need to unpack the idea of performing inferences in virtue of their form without requiring that a thinker must be able to specify the form of the inferences that he or she makes. Still less should it be required that the thinker must offer an explicit account of the form of the inferences as part of his or her reason for making the inferential transitions. My proposal, which is the result of a kind of inference to the best philosophical explanation, is that performing inferences in virtue of their form involves meeting the conditions for *tacit knowledge* of the corresponding inferential rule.

The notion of tacit knowledge that is in play here goes well beyond that of being able to do something but unable to say how one does it.[7] What it is for a rule to be tacitly known, in the sense that figures in Aunty's argument, can be elucidated in terms of a structure in the explanations of causal processes.[8] The causal processes to be considered are transitions between representations. Thus the inputs to, and outputs from, the processes are physical configurations that have semantic properties. For example, the input configurations might represent letter strings, and the output configurations might represent pronunciations. Given such a process, there may be a pattern in the input-output relation when the inputs and outputs are considered under their semantic (representational) descriptions. Thus, for example, it might be that whenever the input configuration represents a string beginning with the letter 'b' the output configuration represents a pronunciation beginning with the sound /B/. In such a case, we can say that the input-output transitions *conform* to a rule about the task domain; in the example, this would be the rule that letter strings beginning with ‘b’ have pronunciations beginning with /B/. But to say that the transitions conform to the rule is not yet to say that the mechanism that mediates those transitions embodies tacit knowledge of that rule. What is required for tacit knowledge of the ‘b’-to-/B/ rule is that the transitions that conform to the rule should have a *common causal explanation*. This condition is met if there is, within the overall transition-mediating mechanism, a component processor or module that operates as a *causal common factor* to mediate all the transitions that instantiate the ‘b’-to-/B/ pattern.

The first step in Aunty’s argument says that thinking requires concept possession and concept possession requires tacit knowledge of rules of inference. The second step then makes use of the point that tacit knowledge of a rule requires syntactic articulation in the physical configurations that are the inputs to the transitions that are explained by the presence of that tacit knowledge.

Consider again the device that takes representations of letter strings as inputs and produces representations of pronunciations as outputs. Although letter strings have internal structure, there is no requirement as yet that the input representations should have internal structure; something unstructured can represent something structured. But suppose that the device embodies tacit knowledge of the ‘b’-to-/B/ rule. The ‘b’-to-/B/ transitions all have a common causal explanation; there is a component processor that operates as a causal common factor to mediate all those transitions. In that case, the various input configurations that represent strings beginning with the letter ‘b’ need to share some property that will engage or activate the ‘b’-to-/B/ component processor. This will be (i) a physical property that (ii) is correlated with the semantic property that these input representations share (that they all represent strings beginning with the letter ‘b’) and (iii) is a determinant of the input configuration’s causal consequences. In short, this property will meet the minimal conditions for being a *syntactic* property.[9]

If the device also embodies tacit knowledge of the ‘i’-to-/I/ rule and the ‘n’-to-/N/ rule, then an input representation of the letter string ‘bin’ must have three syntactic properties. These will be correlated with the three semantic properties of representing a string beginning with the letter ‘b’, representing a string with ‘i’ in the middle, and representing a string ending in ‘n’. In fact, quite generally and quite independently of any consideration of the LOT hypothesis, transition-mediating mechanisms that embody tacit knowledge of rules must have syntactically structured input representations.

We have not taken it as a background assumption that the physical configurations that underpin thoughts should have internal structure. It is consistent with intentional realism or propositional modularity that an occurrence of the thought that 7+5=12 should be underpinned by activation at a single unit (the neural analogue of a light bulb being switched on). But, by the second step of Aunty’s argument, any physical configuration that figures as input to component processors that embody tacit knowledge of rules of inference must be syntactically structured.

Thus, consider an occurrence of a thought in whose content the concept C is a constituent. Suppose that possession of the concept C requires commitment to at least one pattern of inference (where C occurs in the premises)[10] so that, by the first step of Aunty's argument, the thinker meets the conditions for tacit knowledge of the corresponding rule of inference. Then, to the extent that the thought, as it occurs on this occasion, is apt to figure as a premise in an inference of this form, the physical configuration that underpins the thought must have a syntactic property that encodes the occurrence of the concept C in the thought's content. When several concepts are constituents in the content of a thought, and the thought is apt to figure as a premise in inferences of various forms associated with possession of those concepts, the physical configuration corresponding to the thought must have several syntactic properties. These will be physical properties that encode the occurrence of the various concepts as constituents in the thought's content and, indeed, encode the structural relations amongst those constituents. Thus, Aunty's argument takes us from the initial assumption of intentional realism to a version of the LOT hypothesis.

2. RESPONSE TO AN OBJECTION: 'NO NEED TO DESCEND'

Aunty's argument involves a shift between levels, a *downward inference* from the personal to the subpersonal level. We start with personal-level notions of judgement, thought and inference and move to subpersonal-level notions of physical configuration and transition-mediating mechanism, notions that belong in the science of information-processing psychology. Someone who rejects the idea that our conception of ourselves as thinking beings involves a commitment to the truth of the LOT hypothesis may take this shift between levels as the target for an objection. In particular, it may be said that, where a philosophical theory supports such an inference from the personal to the subpersonal level, this is the product of an assumption that the inter-level relation is one of *reduction*. If persons are conceived in purely mechanistic terms then inevitably there will be subpersonal-level requirements (such as the truth of the LOT hypothesis) for personal-level conditions (such as conceptualised thought).

But, according to the imagined objector, putative downward inferences will be undermined once we take proper account of our distinctive nature. We are conscious beings, our thinking is subject to normative requirements, we arrive at judgements and decisions on the basis of reasons, and rationalising explanations of our beliefs and actions in terms of reasons are of a kind quite different from scientific explanations. This distinctive nature cannot be adequately described in the terms favoured by information-processing psychology (nor in the terms of neuroscience, biology, chemistry or physics). If we insist on a non-reductionist conception of ourselves as conscious thinking subjects and agents then we shall see that such requirements as there are on personal-level conditions can be met from the rich resources of the personal level itself. There is *no need to descend* to the subpersonal level of information-processing psychology.[11]

So far, the envisaged objection is entirely schematic. One way of developing it is to argue, first, that a thinker needs to be consciously aware of the forms of the inferential transitions that he performs and, second, that this personal-level requirement undermines the argument for syntactic structure at the subpersonal level. If the form of the inference is already available to the thinker in conscious awareness then why, the objector asks, does it need to be encoded in physical configurations as well?

The objector insists that transitions in thought are subject to norms of rationality. What goes along with this insistence is the idea that if thought is to yield knowledge it must measure up to requirements of rational justification; reliable correlation between thought and reality is not enough. One fairly modest component in a not-purely-reliabilist epistemology is that knowledge requires some appreciation of how one could be getting things right.[12] When a belief is arrived at by an inferential transition that instantiates a valid form, there are some obvious constraints on how this idea is to be elaborated. The thinker cannot be required to use an overarching justificatory principle as an additional premise in his reasoning; such a requirement would be regressive. It would also be wrong to require that the thinker provide a theoretical account of the valid form that his inference instantiates; such a requirement would be too strict. But, it seems plausible that, in order to have an appreciation of how he could be getting things right, the thinker should, in some way, be aware of the form of his inference.

This requirement of *awareness of logical form* goes beyond what is needed for tacit knowledge of rules of inference, for tacit knowledge of rules may be present in a system that lacks consciousness altogether. This is as it should be for a notion of tacit knowledge that is to serve the general purposes of information-processing psychology and cognitive science. But we can agree with the objector that, when we are concerned with knowledge and rationality in conscious thinking subjects, a total lack of awareness is not what is wanted. According to Aunty's argument, the presence of tacit knowledge of rules of inference and the truth of the LOT hypothesis are necessary conditions for conceptualised thought. There is no suggestion that they are sufficient conditions and we should now make it explicit that they are not.

The objector, not satisfied with this concession, maintains that, once the personal-level requirement of awareness of logical form is acknowledged, this actually undermines the downward inference in Aunty's argument. In response, I reject the suggestion that proper attention to our distinctive nature reveals that there is no need to descend to the subpersonal level. Acceptance of downward inferences is consistent with a non-reductionist conception of ourselves.

Aunty's argument begins from a premise about a thinker performing inferential transitions in virtue of their form. The notion of 'in virtue of its form' is, at least in part, causal: the form of an inference is to figure in the causal explanation of a thinker's performing the inferential transition in thought. The two steps of the argument first unpack this causal condition in terms of tacit knowledge and then make use of the general connection between tacit knowledge and syntactic structure. The objector has motivated the claim that the thinker should be aware of the logical form of the inference and I have accepted her claim. This is certainly a personal-level, rather than a subpersonal level, requirement. But clearly, a thinker's being aware of the form of an inference that he performs is not yet sufficient for the form's figuring in the causal explanation of his performing that inference. So the personal-level requirement of awareness of form cannot undermine an argument that aims to show that there are subpersonal level requirements for the personal-level condition of performing inferential transitions in virtue of their form.

The objector may say that I have not treated her objection fairly. She may say that it was no part of her position that performing an inferential transition and being aware of its form can supplant the notion of performing an inference in virtue of its form. Rather, her point was that awareness of form should be incorporated into a personal-level account of performing an inference in virtue of its form and that this personal-level account could then compete with the subpersonal-level account cast in terms of tacit knowledge of a rule of inference. And, she may say, in this competition the personal-level account would be at an advantage since causal transitions whose nature is utterly hidden from the thinker do not measure up to the requirements for rational, knowledge-yielding thought.

If this is the objector's point then there is some justice in it. But it is not correct to suppose that there can be a straightforward competition between a personal-level and a subpersonal-level account of the at-least-partly-causal notion of performing an inference in virtue of its form. After all, the key idea for the first step in Aunty's argument is that the best philosophical elucidation of performing an inference in virtue of its form will reveal that it involves meeting the conditions for tacit knowledge of the corresponding inferential rule. In response to this the objector may say that the conditions for tacit knowledge, cast in terms of transitions of the same form having a common causal explanation, can themselves be met at the personal level, so that there is still no need to descend to the subpersonal-level of information-processing psychology. But what the objector needs to establish is that, even given the background assumption of intentional realism or propositional modularity, the conditions for tacit knowledge could be met at the personal level and not met at the subpersonal level of information-processing psychology. We have been given no reason to suppose that this can be established.

3. THE WORRY ABOUT ELIMINATIVISM: AN INTUITION OF NON-NEGOTIABILITY

Aunty's argument supports a conditional: If we are thinking beings then the LOT hypothesis is true of us; that is, we are LOT beings. Although the argument is relatively non-empirical in character, the question whether we really are LOT beings is a substantive empirical one and answering it requires detailed empirical investigation.[13] Many empirical arguments have been advanced to support the LOT hypothesis but the issue is contested. A good deal of connectionist research aims to show that the kinds of behaviour that have been taken to constitute evidence in favour of the LOT hypothesis may be forthcoming from connectionist networks. Furthermore, it is argued that the networks in question do not meet the conditions for tacit knowledge of rules or syntactically structured representations either at the proprietary level of network description in terms of units and connections, activations and weights, or at any higher level of description. It seems reasonable to allow that it is epistemically possible (whether or not it is likely) that we may turn out not to be LOT beings. But then Aunty's argument would support an eliminativist *modus tollens*. From the premise that we are not LOT beings we would be able to conclude that we are not thinking beings.[14]

Imagine, for a moment, that empirical evidence decisively supported the thesis that we are not LOT beings. It seems that, in those circumstances, we would face a stark choice between two alternatives. On the one hand, we could perform the *modus tollens* inference and cease to regard each other and ourselves as thinking beings. On the other hand, we could conclude that there is something wrong with Aunty's argument. But the first alternative seems rationally to require that we abandon our familiar descriptions of ourselves and others as believing and wanting things, as hoping and fearing things, as engaging in reasoning and planning; and there are powerful intuitions proclaiming that this option is not genuinely available to us. Our everyday engagement in folk psychological practice seems to be philosophically non-negotiable. So, we would be driven to the second alternative. If we found ourselves to be in a disobliging world then we would be bound to reject Aunty's argument. We would have to conclude that the philosophical theories that support the argument are in some way flawed.

But, if this is the conclusion that we should draw if things turned out badly for the LOT hypothesis, then should we not draw that conclusion now? For the credentials of a philosophical and non-empirical argument ought not to depend on what our internal information-processing machinery turns out to be like. Given the intuition of non-negotiability concerning our engagement in folk psychological practice, it appears that the very possibility of an eliminativist *modus tollens* already counts against Aunty's argument. Indeed, that intuition of non-negotiability seems to count against any argument that purports to uncover substantive requirements for thought if there is a prospect that empirical investigation may reveal that those requirements are not met.

However, against all these considerations we must set the fact that a philosophical theory that avoids the threat of eliminativism by imposing no substantive empirical requirements for thought is itself counterintuitive. The threat of eliminativism is posed by any theory that imposes requirements that concern internal information processing. The threat is avoided by theories that impose no requirements that go beyond readily observable patterns in behaviour. But theories that make thought supervene on behavioural trajectory face familiar counterexamples. We can imagine systems that produce the right kind of behaviour but are no more capable of thought than is a puppet or a jukebox.[15]

Standing ready to perform an eliminativist *modus tollens* goes against the non-negotiability of our engagement in folk psychological practice. Restricting our philosophical theories to those that totally avoid the threat of eliminativism has counterintuitive consequences. Neither option is attractive. But in my view we can respect the intuition of non-negotiability even while embracing the philosophical theories that support Aunty's argument.

We can accept that those philosophical theories provide the best elaboration and precisification of our current conception of a thinking being and that Aunty's argument correctly draws out a necessary condition for falling under that conception. But we can also allow that it is part of our current conception that we ourselves are thinking beings: being one of us is a sufficient condition for falling under that conception. Suppose that these claims about a necessary condition and a sufficient condition for falling under our current conception of a thinking being are both correct. It follows that if we are not LOT beings then our current conception dictates both that we are and that we are not thinking beings. In a disobliging world, our current conception of a thinking being would be of no use to us, since it would dictate contradictory answers to the question whether we are thinking beings.

If we turn out not to be LOT beings then we must negotiate our way to a revised conception of what it is to be a thinking being.[16] This conceptual negotiation would proceed under two constraints. The revised conception should be one under which we fall; so it should not involve a commitment to the truth of the LOT hypothesis.[17] And the revised conception should rationally sustain as much as possible of our folk psychological practice.[18] It is by acknowledging this pair of constraints on the process of revision that we honour the intuition of non-negotiability concerning our engagement in folk psychological practice.

In response to the worry about eliminativism, what is being proposed is that the concept of a thinking being has at least two components. There is an exemplar component that specifies sufficient conditions: we, at least, are thinking beings. And there is a more theoretical component which, according to Aunty's argument, imposes a necessary condition: thinking beings are LOT beings. There is no logical guarantee that the items that meet the sufficient conditions also meet the necessary condition and in a disobliging world the two components lead to contradictory verdicts on cases.

This proposal is somewhat analogous to what David Lewis (1997) says about colour concepts. He begins from the thought that our folk theory of colours contains principles linking colours and colour experiences, such as: 'When a red thing is before someone's eyes, it typically causes in him an experience of redness.'[19] If our concepts of colours and of colour experiences are concepts of properties of objects and of inner states that are simultaneously implicitly defined by our folk theory, then a philosophical theory about colours and colour experiences is likely to include such principles as these:

> D1 *Red* is the surface property of things which typically causes experience of red in people who have such things before their eyes.
>
> D2 *Experience of red* is the inner state of people which is the typical effect of having red things before the eyes.[20]

The problem with D1 and D2 is that what they say, while true, does not distinguish the pair <red, experience of red> from other similar pairs such as <green, experience of green>. Something must be added to the folk theory of colour in order to individuate specific colours, such as red. Lewis suggests that there may be different versions of this additional component for different groups of users of the concept, each version specifying relatively parochial examples. Thus, amongst followers of Australian Rules football, it will suffice to say 'that red is the colour of the diagonal stripe on an Essendon Football Club jumper'.[21]

According to the (parochial) exemplar component of the concept of red, being the colour of the Essendon stripe is sufficient for being red: Essendon stripes (at least) are red things. From the theoretical component we can derive a necessary condition for being red: If something is red then there is a type of colour experience and a type of inner state that is typically caused in people who have that thing before their eyes. But there is no logical guarantee that there is a single type of colour experience and a single type of inner state that is typically produced in people by the diagonal stripe on an Essendon jumper. If there is not, then the two components of the concept yield contradictory pronouncements. If the world turns out to be disobliging in this respect then we must negotiate our way to revised, presumably relativised, colour concepts.

In this case, it seems clear that the option of being open to the possibility of conceptual revision in the light of empirical discoveries is preferable to two more extreme options. On the one hand, there is the option of standing ready to abandon colour talk altogether if colour experiences and the corresponding inner states turn out to vary from person to person. On the other hand, there is the option of rejecting any philosophical theory about colours and colour experiences that includes principles such as D1 and D2. My claim is that, similarly in the case of the concept of a thinking being, the option of being open to the possibility of conceptual revision is preferable to the more extreme options of standing ready to perform an eliminativist *modus tollens* or rejecting any philosophical theory that imposes substantive empirical requirements for thought.

4. A SERIOUS PROBLEM: ARMCHAIR KNOWLEDGE

So far, I have responded to the 'no need to descend' objection and addressed the worry about eliminativism. But there is still a serious problem for Aunty's argument.

Suppose that the LOT hypothesis is true and that the concept of a thinking being is in good order. It seems that, by relying on my grasp of the exemplar component of the concept of a thinking being, I can know that I am a thinking being. In fact, it seems that I have more than one way of knowing this. Since at least some thinking is conscious, first-personal awareness of my own conscious mental states also assures me that I am a thinking being. Either way, provided that the LOT hypothesis is in fact true, this knowledge seems to be available to me ahead of any empirical investigation of the information-processing mechanisms inside my head.

By relying on my grasp of the theoretical component of the concept of a thinking being, engaging in some inferences to the best philosophical explanation, and following through Aunty's argument I can, if the argument is a good one, come to know that a thinking being must be an LOT being. I know that if I am a thinking being then I am an LOT being.

Without conducting any detailed empirical investigation, I can have two pieces of knowledge that provide the premises for a simple *modus ponens* inference:

LOT(1) I am a thinking being.
LOT(2) If I am a thinking being then I am an LOT being.
Therefore:
LOT(3) I am an LOT being.

But if, given knowledge of the premises, simply performing the inference yields knowledge of the conclusion then I can know that the LOT hypothesis is true (of me, at least) without any detailed empirical investigation. This seems counterintuitive. If the LOT hypothesis is indeed true then knowledge of that fact will be the result of experiments, computational modelling and, more generally, detailed comparison of the successes and failures of competing research programmes.

Aunty's argument supports a conditional, LOT(2), that gives rise to an instance of *the problem of armchair knowledge*. Both LOT(1) and LOT(2) can be known from the armchair, yet knowledge of LOT(3) surely requires an investigative methodology rather than an armchair methodology.

Another instance of the problem of armchair knowledge arises from my proposal (following Lewis) about colour concepts. Suppose that our colour concepts are in good order and consider the following *modus ponens* inference:

> RED(1) This [pointing at the diagonal stripe on an Essendon jumper] is red.
> RED(2) If this is red then there is a type of colour experience and a type of inner state that is typically caused in people who have this before their eyes.
> Therefore:
> RED(3) There is a type of colour experience and a type of inner state that is typically caused in people who have this before their eyes.

By relying on my mastery of the exemplar component of the concept of red (the parochial exemplar component that applies to my group), I can know that this Essendon stripe is red. Indeed, I have more than one way of knowing this since I can often know what colour something is just by looking at it. Having seen many Essendon jumpers, I can recognise this item as being the colour of the Essendon diagonal stripe. Either way, knowledge of RED(1) is available to me ahead of any investigation of other people's colour experiences or inner states. By relying on my grasp of the theoretical component of the concept of red (including the principles D1 and D2), I can know that if something is red then there is a type of colour experience and a type of inner state that is typically caused in people who have that thing before their eyes. So I can know RED(2). But it is implausible that, without rising from the armchair save perhaps to look at an Essendon football jumper, I can know the conclusion RED(3).

The same kind of problem also arises from externalist theories about the intentional content of mental states. Sitting in my armchair, I can know that I am thinking that water is wet. Sitting in my armchair and following through some of the arguments in externalist theories, I can, if those theories are correct, know that if I am thinking that water is wet then I am, or have been, in an environment containing samples of water. Sitting in my armchair, I can draw the obvious conclusion. Yet it is counterintuitive that a mere exercise in armchair reflection should provide me with knowledge that I am, or have been, in a watery environment.[22]

The problem of armchair knowledge is serious and it may be thought to cast doubt on the philosophical theories that give rise to it and, in particular, on the theories that support Aunty's argument for the language of thought. But in my view the problem is soluble. The key to the solution is that armchair epistemic warrants for the premises of an argument are not always transmitted to the conclusion of the argument, even when the inference is palpably valid. There are limitations on the transmission of epistemic warrant from premises to conclusion.[23]

5. A SOLUTION: LIMITATIONS ON TRANSMISSION OF EPISTEMIC WARRANT

My proposals for principles that limit the transmission of warrant are motivated by the claim that failure of transmission of epistemic warrant is the analogue, within the thought of a single subject, of the dialectical phenomenon of begging the question. In order to make use of this claim, we need to have an account of begging the question and for this I rely on Frank Jackson (1987). He says that an argument begs the question when 'anyone – or anyone sane – who doubted the conclusion would have background beliefs relative to which the evidence for the premises would be no evidence'.[24]

According to Jackson's view of what is achieved by advancing an argument for a conclusion, the speaker invites the hearer to borrow evidence, or other considerations, in favour of the premises of the argument. By her choice of premises she provides an indication as to what kind of considerations these are. Typically, evidence counts in favour of a proposition only relative to particular background assumptions and often the relevant background assumptions are shared between speaker and hearer. But when background assumptions are not shared, it is possible that the considerations that count in favour of the premises relative to the speaker's background assumptions do not count in favour of the premises relative to the hearer's background assumptions. Suppose that a speaker sets out to convince a doubting hearer of the truth of some conclusion. The speaker begs the question against the hearer if the hearer's doubt rationally requires him to adopt background assumptions relative to which the considerations that are supposed to support the speaker's premises no longer provide that support. A question-begging argument 'could be of no use in convincing doubters'.[25]

Convincing a doubter and providing an epistemic warrant both involve ruling out various ways in which a proposition could have been false. A speaker's evidence for her premises rules out various ways in which those premises could have been false, ways left open by the speaker's background assumptions. A hearer who doubts the conclusion of the argument may have background assumptions that leave a wider range of possibilities open, and the speaker's evidence for the premises may not rule all those possibilities out. Indeed, the speaker's evidence may leave untouched the ways in which, according to the hearer, the conclusion could be false.

In a similar way, a thinker's epistemic warrants for believing the premises of an argument rule out various ways in which those premises could have been false, ways left open by background assumptions that the thinker is, in that context, epistemically entitled to make. But it does not follow that those same warrants rule out ways in which the conclusion could have been false which are left open by assumptions that the thinker is entitled to make. It may be, for example, that in the context of considering the conclusion the thinker is entitled to make fewer assumptions than when only the premises were under consideration. In that case, epistemically adequate warrants for the premises may be inadequate to rule out the possibilities that now need ruling out. On the other hand, it may be that the assumptions that the thinker is entitled to make already leave open no ways in which the conclusion could have been false. In that case, the thinker is epistemically entitled to believe the conclusion but the epistemic warrants for believing the premises contribute nothing to that entitlement. In either case, the thinker's epistemic warrants for believing the premises do not add up to an epistemically adequate warrant for the conclusion.

The analogy between convincing a doubter and providing an epistemic warrant motivates the following principle:

> **First Limitation Principle**
> Epistemic warrant cannot be transmitted from the premises of a valid argument to its conclusion if, for one of the premises, the warrant for that premise counts as a warrant only against the background of certain assumptions and acceptance of those assumptions cannot be rationally combined with doubt about the truth of the conclusion.

In order to apply this principle to any particular argument, we need to identify assumptions such that, for one of the premises, it is only against the background of those assumptions that the epistemic warrant for the premise counts as a warrant. Then we need to show that acceptance of those assumptions cannot be combined with doubt about the truth of the conclusion. There do seem to be cases that fit this pattern.[26] But it is not clear how we can apply the principle to explain the failure of transmission of warrant from LOT(1) and LOT(2) to LOT(3).

In the case of the conditional premise LOT(2), it might reasonably be said that it is only against the background of the assumption of intentional realism that the premise is supported by neo-Fregean philosophical theory. But it is surely not true that acceptance of this assumption cannot be rationally combined with doubt about the conclusion LOT(3). In the case of the first premise, LOT(1), the epistemic warrant is constituted either by grasp of the exemplar component of the concept of a thinking being or else by awareness of my own conscious mental states. But in neither case is there an obvious candidate for the role of background assumption without which the epistemic warrant would not count as a warrant.

There is, however, a very basic assumption that lies in the background of any epistemic project; namely, the assumption that there is such a proposition as the proposition for which one is attempting to provide evidence, justification or warrant. The notion of a proposition that figures in this assumption is not to be construed in a metaphysically committed way. If a thinker is attempting to provide a warrant for believing A then the basic background assumption is simply that there is such a thing to think as A. If there were no such thing to think as A then there could be no question of anything constituting an epistemically adequate warrant for believing A. So we can make explicit a second principle that is arguably a consequence of the first:

> **Second Limitation Principle**
> Epistemic warrant cannot be transmitted from the premises of a valid argument to its conclusion if, for one of the premises, acceptance of the assumption that there is such a proposition for the knower to think as that premise cannot be rationally combined with doubt about the truth of the conclusion.

One way in which the assumption that figures in this principle could turn out to be false would be if one of the purported conceptual constituents in the premise were revealed to be internally incoherent, dictating contradictory answers to the question whether some particular item falls under the concept. To that extent, the principle holds some promise of providing a solution to the instance of the problem of armchair knowledge that is posed by Aunty's argument. But, on the other hand, it is clear that acceptance of the assumption that there is such a thing to think as that I am a thinking being and, in particular, acceptance of the assumption that the concept of a thinking being is in good order can be rationally combined with doubt about the truth of the LOT hypothesis. It is only the acceptance of Aunty's argument that generates rational tension between acceptance of the background assumption and doubt about the conclusion.

It is clear what kind of modification of the principle is required if it is to provide a solution to the problem of armchair knowledge that arises from Aunty's argument and the modification is not merely opportunistic or *ad hoc*. To see this, we need to return to begging the question and focus on the fact that arguments may have several premises. Suppose that a speaker advances a multi-premise argument in an attempt to convince a hearer who doubts that argument's conclusion. The speaker offers various considerations for borrowing; they are considerations that count in favour of the premises relative to the speaker's background assumptions. If the hearer's doubt by itself rationally requires him to adopt background assumptions relative to which one of the speaker's premises is no longer supported by the considerations that she offers for borrowing then the speaker begs the question against the hearer. That is the kind of case that Jackson describes.

But there is a more complicated scenario in which it is no less true that the argument, as advanced by the speaker, will be of no use in convincing the doubting hearer. If the hearer is to be convinced then he must accept the considerations that the speaker offers in support of her premises. In addition, he must not differ from the speaker in his background assumptions in such a way that the premises are no longer supported by those considerations. Suppose that the hearer's doubt about the conclusion, when put together with acceptance of the considerations that the speaker offers in support of some of the premises, rationally requires him to adopt background assumptions relative to which some other one of the premises is no longer supported by the considerations offered in support of it. That is enough to ensure that the argument, as advanced by the speaker, will be of no use in convincing the hearer. So, if failure of transmission of epistemic warrant is the analogue, within the thought of a single subject, of the dialectical phenomenon of begging the question then we should expect the following pair of limitation principles, of which the second is arguably a consequence of the first:

> **First Limitation Principle (generalised version)**
> Epistemic warrant cannot be transmitted from the premises of a valid argument to its conclusion if, for one of the premises, the warrant for that premise counts as a warrant only against the background of certain assumptions and acceptance (i) of those assumptions and (ii) of the warrants for the other premises cannot be rationally combined with doubt about the truth of the conclusion.
>
> **Second Limitation Principle (generalised version)**
> Epistemic warrant cannot be transmitted from the premises of a valid argument to its conclusion if, for one of the premises, acceptance (i) of the assumption that there is such a proposition for the knower to think as that premise and (ii) of the warrants for the other premises cannot be rationally combined with doubt about the truth of the conclusion.

It is this last principle that provides a solution to the instance of the problem of armchair knowledge that is posed by Aunty's argument.

Suppose that a thinker accepts that there is such a thing to think as the premise LOT(1), that he himself is a thinking being. Suppose, in particular, that he accepts that there is no internal incoherence, no source of contradictions, in the concept of a thinking being. In that case, the thinker must accept the assumption that the items, such as himself, that meet the sufficient conditions for falling under the concept also meet the necessary conditions. Acceptance of that assumption does not, by itself, rationally preclude doubt about whether he himself is an LOT being. But suppose, in addition, that the thinker accepts the epistemic warrants for the premises LOT(1) and LOT(2). The epistemic warrant for believing LOT(1) is provided either by the exemplar component of the concept of a thinking being or else by his awareness of his own conscious mental states. But it is the warrant for believing the conditional premise LOT(2) that figures crucially in the solution to the problem of armchair knowledge. That warrant is provided by a battery of philosophical theory and by Aunty's argument. Acceptance of the assumption that the items that meet the sufficient conditions for falling under the concept of a thinking being also meet the necessary conditions *and of the warrant for LOT(2)* cannot be rationally combined with doubt about whether the thinker himself is an LOT being. So the limitation principle is triggered and epistemic warrant cannot be transmitted from LOT(1) and LOT(2) to the conclusion LOT(3). The warrants for believing the premises do not add up to an epistemically adequate warrant for the conclusion.

Because there are limitations on the transmission of epistemic warrant, armchair knowledge of the premises LOT(1) and LOT(2) does not lead to armchair knowledge of the conclusion LOT(3). Aunty's argument does not, after all, have the counterintuitive consequence that I can know that the LOT hypothesis is true (of me, at least) without any detailed empirical investigation. In a similar way we can provide a solution to the problem of armchair knowledge that is posed by my proposal about colour concepts.

CONCLUSION

I began by reviewing Aunty's own argument for the language of thought and have then responded to the 'no need to descend' objection, addressed the worry about eliminativism, and proposed a solution to the problem of armchair knowledge.

It is plausible that the worry about eliminativism provides some of the motivation for the 'no need to descend' objection. The downward inference from the personal to the subpersonal level seems particularly problematic to the extent that it poses a threat of eliminativism. I have argued that we do not have to choose between, on the one hand, acknowledging that in a disobliging world we should be rationally bound to abandon our folk psychological practice and, on the other hand, rejecting all such downward inferences. There is a third option. We should be open to the possibility of conceptual revision in the light of empirical discoveries. If I am right about this then the motivation for the 'no need to descend' objection is reduced. But we should retain the idea that, for conceptualised thought, there is a personal-level requirement of awareness of logical form in addition to the subpersonal-level requirements of the presence of tacit knowledge of rules of inference and the truth of the LOT hypothesis.

The worry about eliminativism can be addressed. But it points towards a genuine and serious problem, namely, that Aunty's argument seems to offer a route to armchair knowledge about the workings of the information-processing machinery inside our heads. Furthermore, our solution to the problem of armchair knowledge depends crucially on a concession that was made to the worry about eliminativism. We respect the intuition of non-negotiability concerning our engagement in folk psychological practice by acknowledging a pair of constraints on conceptual negotiation. A revised conception of a thinking being should be a conception under which we fall and it should rationally sustain as much as possible of our folk psychological practice. This pair of constraints reflects the idea that, along with a theoretical component, the concept of a thinking being has an exemplar component: we, at least, are thinking beings. Without the proposal that the concept has relatively independent exemplar and theoretical components, the solution to the problem of armchair knowledge would not work. The solution depends on the fact that acceptance of a background assumption about conceptual coherence cannot be rationally combined with a particular empirical doubt. And it is only because of the independence of its two components that the concept of a thinking being lives at empirical risk of being torn apart.

Martin Davies
Australian National University (Canberra)
Australia

NOTES

[1] Davies, 1992.
[2] Fodor, 1987, p. 139: 'It turns out that dear Aunty is, of all things, a New Connectionist Groupie.'
[3] Cf. Rey, 1995.
[4] See Ramsey, Stich and Garon, 1990.
[5] The objection, the worry, and the problem arise even if we consider, instead of the language of thought hypothesis, the conditional statement: If intentional realism is true then the LOT hypothesis is true.
[6] In the original paper, I developed this step of the argument in two slightly different ways, drawing in turn on Evans, 1982, and on Peacocke, 1990. Here I follow the latter version. Peacocke's notion of concept possession is presented in Peacocke, 1992, chapter 1; see p. 6 for the claim that certain inferences should be found primitively compelling in virtue of their form. In that book, Peacocke proposes that concepts can actually be individuated by their possession conditions, where those conditions are specified in terms of something like a functional (inferential) role. But this is not essential to Aunty's argument.
[7] Polanyi, 1967.
[8] The account owes much to Evans, 1981. See Davies, 1987, for some details omitted here and Davies, 1995, for application of the notion to connectionist networks.
[9] See Fodor, 1987, pp. 16–21.
[10] The argument presented in the text would not apply if the only rules of inference associated with the concept C were introduction rules.
[11] This objection is considered at greater length in Davies, 2000a.

[12] See Brewer, 1995, 1996, 1999.
[13] This is so even given the assumption that intentional realism is true of us.
[14] Cf. Ramsey, Stich and Garon, 1990, p. 500: '*If* connectionist hypotheses [of a particular sort] turn out to be right, so too will eliminativism about propositional attitudes.'
[15] See Peacocke, 1983; Block, 1981, 1995.
[16] The process of revision will be informed by the particular ways in which the world turns out to be disobliging.
[17] On the assumption that the philosophical theories supporting Aunty's argument do provide the best elaboration and precisification of our current conception, we need to revise that conception in order to avoid a commitment to the truth of the LOT hypothesis.
[18] We would not abandon the idea that we engage in deductive inference; but we would, presumably, adjust our conception of what is involved in accepting or performing an inference in virtue of its form.
[19] Lewis, 1997, p. 326.
[20] Ibid., p. 327.
[21] Ibid., p. 335.
[22] See Davies, 2000b.
[23] For an earlier development of this idea, see Davies, 1998.
[24] Jackson, 1987, p. 111.
[25] Ibid., p. 112.
[26] See the discussion of Moore's anti-sceptical argument in Davies 1998, 2000b; and cf.Wright, 1985.

REFERENCES

Block, N. 1981: Psychologism and behaviorism. *Philosophical Review*, 90, 5–43.

Block, N. 1995: The mind as the software of the brain. In D.N. Osherson and E.E. Smith (eds), *An Invitation to Cognitive Science, Volume 3: Thinking*. Cambridge, MA: MIT Press, 377–425.

Brewer, B. 1995: Compulsion by reason. *Proceedings of the Aristotelian Society, Supplementary Volume* 64, 237–53.

Brewer, B. 1996: Internalism and perceptual knowledge. *European Journal of Philosophy*, 4, 259–75.

Brewer, B. 1999: *Perception and Reason*. Oxford: Oxford University Press.

Chomsky, N. 1965: *Aspects of the Theory of Syntax*. Cambridge, MA: MIT Press.

Davies, M. 1987: Tacit knowledge and semantic theory. Can a five per cent difference matter? *Mind*, 96, 441–62.

Davies, M. 1992: Aunty's own argument for the language of thought. In J. Ezquerro and J.M. Larrazabal (eds), *Cognition, Semantics and Philosophy:*

Proceedings of the First International Colloquium on Cognitive Science. Dordrecht: Kluwer Academic Publishers, 235–71.
Davies, M. 1995: Two notions of implicit rules. In J.E. Tomberlin (ed.), *Philosophical Perspectives 9: AI, Connectionism and Philosophical Psychology*. Atascadero, CA: Ridgeview Publishing Company, 153–83.
Davies, M. 1998: Externalism, architecturalism, and epistemic warrant. In C. Wright, B.C. Smith and C. Macdonald (eds), *Knowing Our Own Minds*. Oxford: Oxford University Press, 321–61.
Davies, M. 2000a: Persons and their underpinnings. *Philosophical Explorations*, 3, 43–62.
Davies, M. 2000b: Externalism and armchair knowledge. In P. Boghossian and C. Peacocke (eds), *New Essays on the A Priori*. Oxford: Oxford University Press, 384-414.
Evans, G. 1981: Semantic theory and tacit knowledge. In S. Holtzman and C. Leich (eds), *Wittgenstein: To Follow a Rule*. London: Routledge and Kegan Paul, 118–37. Reprinted in *Collected Papers*. Oxford: Oxford University Press, 1985.
Evans, G. 1982: *The Varieties of Reference*. Oxford: Oxford University Press.
Fodor, J. 1985: Fodor's guide to mental representation. *Mind*, 94, 55–97. Reprinted in *A Theory of Content and Other Essays*. Cambridge, MA: MIT Press, 1990.
Fodor, J. 1987: *Psychosemantics*. Cambridge, MA: MIT Press.
Jackson, F.C. 1987: *Conditionals*. Oxford: Blackwell Publishers.
Lewis, D. 1997: Naming the colours. *Australasian Journal of Philosophy*, 75, 323–42.
Peacocke, C. 1983: *Sense and Content*. Oxford: Oxford University Press.
Peacocke, C. 1990: Content and norms in a natural world. In E. Villanueva (ed.), *Information, Semantics and Epistemology*. Oxford: Basil Blackwell, 57–76.
Peacocke, C. 1992: *A Study of Concepts*. Cambridge, MA: MIT Press.
Polanyi, M. 1967: *The Tacit Dimension*. London: Routledge and Kegan Paul.
Ramsey, W., Stich, S. and Garon, J. 1990: Connectionism, eliminativism and the future of folk psychology. In J.E. Tomberlin (ed.), *Philosophical Perspectives 4: Action Theory and Philosophy of Mind*. Atascadero, CA: Ridgeview Publishing Company, 499–533.
Rey, G. 1995: A not 'merely empirical' argument for a language of thought. In J.E. Tomberlin (ed.), *Philosophical Perspectives 9: AI, Connectionism and Philosophical Psychology*. Atascadero, CA: Ridgeview Publishing Company, 201–22.
Wright, C. 1985: Facts and certainty. *Proceedings of the British Academy*, 71, 429–72.

PAUL EKMAN

EMOTIONAL AND CONVERSATIONAL NONVERBAL SIGNALS

In this paper I revise and expand formulations which distinguished among a number of different types of body movements and facial expressions (Ekman and Friesen 1969). Some of the terminology and most of the conceptual distinctions have been preserved, but refinements and expansions have benefited from empirical findings and theoretical developments. The crucial issue remains to distinguish among quite different activities, which are shown in facial and/or bodily movement, but which have quite different functions, origins and coding.

1. EMBLEMS

Emblems are the only true 'body language', in that these movements have a set of precise meanings, which are understood by all members of a culture or subculture. The term 'emblem' I borrowed from Efron (1968), the pioneer in studying cultural differences in body movements. Emblems are socially learned and thus, like language, culturally variable. A message may have an emblem in one culture, and no emblem in another cultural setting. Or the same movement pattern may have quite different meanings in different cultural settings.

There are, however, multicultural emblems, which may occur for quite different reasons. First, and most obvious, some emblems from one culture may be adopted by members of another culture who have observed them. The 'finger' emblem in common use in North America is well known by inter-cultural contact, and sometimes used in other cultures. Such contact may be direct, or through mass media. Darwin (1998) proposed quite a different mechanism to explain his observation that the shrug, which denotes helplessness, while not universal, is quite widespread. It is a movement, he said, which is antithetical to the movement patterns used to denote the capability to attack. If a culture were to develop an emblem for helplessness, then, from Darwin's reasoning, it would likely be this antithetical movement. Eibl-Eibesfeldt (1970) has claimed that the eyebrow flash to denote a greeting is universal. I disagree; there are some

Jesus M. Larrazabal and Luis A. Pérez Miranda (eds.), Language, Knowledge, and Representation: 39-50.

cultures where it is not in use (e.g. the United States). But it is quite widespread, appearing in many cultures. I explain its frequent occurrence as due to the selection of one element — raising of the eyebrows – of the full display of surprise. If a culture develops an emblem for an initial greeting, and uses the face in addition or in place of the hands, then it is likely that part of a surprise display would be used.

Emblems may repeat a word as it is said, replace a word in a flow of speech, provide a separate comment related to the words spoken, or occur as the sole reply. Emblems may be iconic, in which the movements look in some way like the message they are signifying, or arbitrarily coded. Emblems most often involve the hands, but some are performed using the shoulders, changes in head positioning, or facial movements.

Emblems are typically performed in a 'presentation position': directly in front of the performer, between the waist and the head. They usually have a staccato or punctuated appearance, with a sudden beginning and ending. The performer is as aware of using an emblem as he or she is of the words being spoken.

An exception is the 'emblematic fragment'. These are the gestural equivalents of slips of the tongue. They occur outside of the presentation position, and only part of the full emblem is performed, in a non-punctuated fashion. The onset and offset may be gradual rather than abrupt. And the performer usually is unaware of making the movements. Like verbal slips, emblematic fragments may reveal repressed information, or deliberately suppressed information (Ekman 2001).

Emblems can be performed with hand, head, or facial movements. In those cultures we have studied, the majority of the emblems are performed with hand movements. Emblems are of special use when people can not rely on words (too much noise, distance, etc.). When used with speech they can either add a second layer to the conversation, or emphasize and make the spoken words more interesting.

Elsewhere (Ekman and Friesen 1972; Johnson *et al.* 1975) we have described the method we have used to survey the repertoire of emblems in different cultural settings. This method has been used in the United States (Johnson *et al.* 1975), Iran (Trupin unpublished dissertation 1976), Israel (Broide unpublished masters thesis 1977) and Japan and Papua New Guinea (the last two are unpublished findings from my own research). These cultures differed in the sheer number of emblems (67 in the U.S. to over 200 in Israel), and in the types of messages for which there is an emblem.

It is my impression that it is easy to learn foreign emblems, easier than it is to learn foreign words. But I have done no research to support that casual observation.

Although emblems resemble spoken language, they do not appear to have syntax. Sometimes a person will use a string of emblems, in which the sequence denotes the set of messages in the intended order, but such sequences are not common, and no other more complex relationship among emblems themselves has been evident.

2. ILLUSTRATORS

We (Ekman and Friesen 1969) coined this term to refer to movements that illustrate speech. They are intimately related to the speaker's speech on a moment-to-moment basis, usually augmenting what is said, but sometimes contradicting it. Although Efron did not use this term, he described five different types of illustrators, and I preserve his terminology for naming each type, and add two more to his list (6 and 7 below):

1. Batons time out, accent or emphasize a particular word or phrase, 'beat out the tempo of mental locomotion'.
2. Ideographs sketch a path or directing of thought, 'tracing the itinerary of a logical journey'.
3. Deictic movements point to a present object.
4. Kinetographs depict a bodily action.
5. Spatial movements depict a spatial relationship.
6. Pictographs draw a picture of their referent.
7. Rhythmic movements depict the rhythm or pacing of an event.

These illustrators are typically performed with the hands, although the head may be involved, or even the feet. I made these distinctions (Ekman and Friesen 1969; Ekman and Friesen 1972) when I had been studying bodily movement. After another decade of measuring facial muscular movement, I noticed that certain facial movements can be used as batons, and that this category can be subdivided (Ekman 1979). There are facial (as well as bodily) batons that accent a word. These baton-accents typically co-occur with an increase in loudness. There are also baton-underliners that emphasize a phrase. We have found it is nearly impossible to deliberately disconnect the change in loudness and the baton-accent. We have not found anyone who, within a phrase, can deliberately increase loudness on one word and place the baton-accent on another word, without unintentionally increasing the loudness of the baton-accented word.

Nearly all of the facial batons involve either brow raising or brow lowering, although it is possible to use nearly any facial movement for this purpose. Brow lowering and brow raising are highly visible facial movements, and movements that we have found are very easy to perform, even by young children (Ekman *et al.* 1980b). Although some individuals utilize raising more than lowering, or vice versa, for most people the facial movement deployed as a baton-accent is related to the content of the word being emphasized. With words such as 'easy, light, good, etc.' a brow raise is used, while brow lower is used to emphasize words such as 'difficult, dark, bad', etc. If a person tries to use brow lower to emphasize a word such as 'light' or 'good', the voice sounds strange, and the performance seems rough. I offered two non-exclusive explanations for the association of brow lower and brow raise with negative and positive words:

[B]row lowering which is employed in a variety of negative emotions (fear, sadness, distress, anger) should carry an implication of something negative, whereas brow raising would be more likely to suggest surprise or interest. Alternatively, the role played by these two actions in conversational signals may be selected on the basis of their current biological function: brow raise increasing and brow lowering decreasing visual input. Their role in conversational signals would thus be viewed as analogues to their biological adaptive value. Either possibility could be true (Ekman 1979, p. 201).

In unpublished research with Linda Camras (described in more detail in Ekman 1979) we found that brow lowering and raising is also used in quite different ways in statements which ask a question. If the speaker is going to use a brow movement at all, then it is likely that the brows will be raised if the speaker knows the answer to the question, and lowered if the speaker does not know the answer to the question be asked. One of the editors of this volume (L.M.) noted how this contrasts with the use of brows in signing. (In many sign languages, American Sign Language amongst them, raised brows are part of the signal for yes-no questions; lowered brows are part of the signal for wh-questions. However, raised brows are also part of the signal for rhetorical questions, which is in keeping with our findings for non-signers.) The tone of voice and pitch level will also differ with these two types of questions.

Illustrators are socially learned, presumably when language itself is learned. Efron dramatically demonstrated that immigrant Lithuanian Jews and Sicilians use very different types of illustrators, but these differences were not preserved in their offspring who assimilated into the mainstream New York City culture.

Although individuals differ in both the type and frequency with which they use illustrators, there are also individual variations. Illustrators increase with the speaker's involvement with the process of speaking, and with both positive and negative affect, and decrease with lowered involvement, boredom or fatigue. A decrease in illustrators also occurs when a person is carefully weighing each word before it is spoken. Although this may occur with caution or ambivalence, it also has proven to be a useful indicator of deception (Ekman *et al.* 1976). We have also found that illustrators increased when depressed patients recovered (Ekman and Friesen 1974).

Illustrators help explain what is being said verbally. They also can serve a self-priming function, helping the speaker get going or get through a difficult to explain thought. Illustrators command the listener's attention, and can help hold the floor for the speaker. We have found (Ekman *et al.* 1980a) that the use of illustrators impresses others as a sign of sociability and friendliness. I expect students would retain more information from a lecturer who uses illustrators, but I don't know of research to support that speculation.

3. MANIPULATORS

Originally we (Ekman and Friesen 1969) called these movements 'adapters', but I now prefer the more descriptive term 'manipulators'. In these movements one part of the body or face manipulates in some fashion – stroking, pressing, scratching, licking, biting, sucking, etc. – another part of the body or face. An object may also be the object of this attention, or used to perform the manipulation.

The frequency with which these movements occur is amazing, once one begins to notice them. Not just in private, but also in public, people usually touch themselves. One of my favorite classroom exercises is to ask the students to put everything down and stop touching themselves. It is hard for them to maintain this for even five minutes!

While some of these manipulators appear to accomplish grooming or cleaning, many of them seem to have no instrumental goal. Perhaps reassurance or comforting is a possibility, but many manipulators seem simply to reflect a nervousness, or habitual activity. Manipulators appear to be performed on the edge of awareness, in that a person if asked what he or she just did can usually describe the activity, but was not focusing on it as it occurred. Our observations suggest that most people disattend when another person engages in a particularly noticeable, presumably taboo, manipulator such as ear or nose cleaning. The disattending, I believe, also occurs with little awareness.

Individuals differ not only in their favored manipulator but also in the frequency with which they show these behaviors. They often increase with increasing discomfort, although some people show a decrease, and freeze into a tense restrained position when uncomfortable. And, manipulators may increase when people are totally comfortable, with friends, not worrying at all about appearances.

Our research (Ekman *et al.* 1980a; Ekman 1992a) found that others distrust people who show many manipulators; they are commonly interpreted as signs someone is lying. But, in fact, they are not a reliable sign of lying for most people.

4. REGULATORS

We coined this term to draw attention

> ...to actions which maintain and regulate the back-and-forth nature of speaking and listening between two or more interactants. They tell the speaker to continue, repeat, elaborate, hurry up, become more interesting, less salacious, give the other a chance to talk, etc. They tell the listener to pay special attention, to wait just a minute more, to talk, etc. (Ekman and Friesen 1969, p. 82)

While illustrators and emblems (and emotional expressions which I describe next) influence and may also be said to regulate the flow of conversation, I use the term 'regulator' to refer to when those or other actions are primarily functioning for just this purpose alone. Scheflen (1963, 1964, 1965) wrote about regulators, although that is not what he called them, and made many important contributions such as the role of postural shifts and mirroring. I have not focused much on regulators, and my observations are limited.

'Listener responses' were first described, I believe, by Dittman (1972). He focused on a variety of movements that lead the speaker to continue with what he is saying. These may include head nods, agreement-smiles, forward leans, brow raises in exclamation, and so forth. I called these 'agreement listener responses' to distinguish them from another set, 'calls for information' (Ekman 1979). The listener may lower the brow to signal a lack of understanding or puzzlement with what has been said. A brow raise may indicate disbelief, or that what has been said is incredible.

I described earlier but did not before name, 'floor holders', responses made by the speaker to prevent interruptions. I have seen this done by holding a hand out, like a traffic policeman, to prevent the listener from entering the conversation just yet. 'Turn seekers' are the listeners attempts to gain the floor, which may involve such diverse actions as leaning forward, almost rising from a chair, beginning to make the lip movements for speaking a word, etc.

5. EMOTIONAL EXPRESSIONS

We originally used the term 'affect display', but each of those words carries surplus meaning that I would rather avoid. 'Affect' implies more than emotion; it encompasses a broader range of phenomena, such as moods. And 'display' is no better a word than 'expression' which I prefer in deference to Darwin. Also the use of the term 'expression' serves to raise the issue, which has bothered many ethologists, of whether these movements on the face are communicative actions or expressions. This is a false and misleading dichotomy. I maintain (Ekman 1997) that emotional expressions are involuntary signals which provide important information to others. These expressions have been selected and refined over the course of evolution for their role in social communication. As involuntary signals they may occur in response to anything that calls forth an emotion, which may include non-personal events such as a beautiful sunset or thunder, and may be manifest when the individual is alone. The presence of others enhances expressions, as emotions themselves evolved primarily to deal with fundamental life tasks involving child care, mating, dealing with predators and rivals, etc. (Ekman 1992).

It is reasonable to call these signals 'expressions' because they are part of an emotion, they are a sign that an emotion is occurring. I have maintained that a hallmark of an emotion is that it has a signal, in face and/or voice and/or bodily movement. There is no involuntary signal which informs conspecifics what the person is thinking; thoughts are private, but emotions are not. Elsewhere (Ekman 1992) I have described nine characteristics which distinguish emotions from reflexes, moods, affective traits and affective disorders. I think it is a reasonable assumption that any state which shares the characteristics I describe for an emotion will have a signal. But that is only an assumption, not sacred theory, waiting to be challenged by evidence of a secret emotion which has all the other characteristics of the signal emotions.

We do not know much about what people actually derive when they see or hear an expression in the course of a conversation, competing with other sources of information, and embedded in one or another context of previous and simultaneous behaviors, and various expectations. From the research that has been done by removing an expression from context and asking people who are not involved in the situation to make judgments, we know that emotional expressions *can* provide information about the antecedent events, and about single emotion terms. I have proposed that emotional expressions also may convey information about: likely next actions, likely thoughts and plans, the internal state of the expresser which may be in terms of a metaphor, or what the observer might be thinking it would be wise to do next (Ekman 1993).

While the emotional expressions are not learned, we do learn what we termed 'display rules', to manage expressions. These rules which are to some extent individually and culturally variable, specify who can show which emotion to whom and when. Individuals differ in their success in inhibiting, substituting, masking or magnifying their expressions.

My own evidence and the evidence of many others support Darwin's claim that these expressions are universal to our species, and some of them shared with other species. Social constructionists, cultural relativists, and those focused on the language of emotions still argue with this claim (see Ekman 1999 for my most recent review of the evidence and the challenges to that evidence).

Words are not emotions, but representations of emotion. The fact that a language does not have a word for an emotion does not mean the emotion does not exist for the users of that language, but that it has not been labeled with a single word. English speakers can enjoy the suffering of their enemies but they have no single term for it, while Germans have the word *schadenfreude*. Perhaps the mistaken belief that words are emotions, or the most important feature of an emotion, arose because most of the cross cultural research (but not all, see Ekman 1972), asked people in different cultures to choose a single emotion word for each face they were shown. While this did produce very strong agreement across cultures, it was not perfect, for single words do not translate exactly from one language to another. Another technique, still involving words, was to use a story to convey a social context, and ask people to indicate which face fit which social context. Again, very high agreement was observed.

The face is one of the primary sites of emotional expressions; the voice is the other. I believe posture, the positioning of the head and body, is also recruited into the signal of some of the emotions. We (Ekman and Friesen 1978) developed a fine-grained measurement technique for comprehensively describing any facial movement in terms of the particular muscular actions produced. The Facial Action Coding System (FACS) can be used to describe any observed facial action, not just those which may be relevant to emotion. We have found even in highly emotional circumstances that most of the facial movements are not emotional signals, but facial illustrators, primarily batons. A few hundred people have learned FACS and are using it to study pain, infant development, sign language, psychopathology, and psychotherapy, as well as emotion (see Ekman and Rosenberg 1997 for a representative sample of those studies).

There is not just one but a family of facial expressions for each emotion. The family includes variations related to intensity, efforts to control the expression, and perhaps also the particular form of the emotion. Taking anger as an example, there is evidence about intensity variations, from annoyance to rage, and how to distinguish controlled from uncontrolled anger, but we do not know if there are different expressions for self-righteous anger as compared to indignant anger.

It is not certain how many universal expressions there are. The evidence from the judgment of facial expressions is strong for: anger, fear, disgust, contempt, surprise sadness/distress, and enjoyment. Keltner (1995) suggests there may also be a universal expression for embarrassment. The other emotions can be signaled by a face in an instant, a frozen movement, but embarrassment apparently requires the unfolding of a sequence of facial and bodily movements. I suspect there is also an expression for awe, but no one has described it. (The evidence from the study of the voice (e.g., Banse and Scherer 1996) has to date not identified any other emotions which have a unique cross-cultural signal).

I have suggested (Ekman 1992b) that there is an unhappiness group of emotions which shares a particular facial expression, and a happiness group which does so as well. Members of the unhappiness group are sadness, the more agitated version of sadness called distress, guilt, shame, discouragement, and disappointment. My claim is that their appearance on the face is not much different, with only minor variations for each of these related states. Perhaps the voice might provide separate signals for these presumed emotions, but research to determine this has not yet been done.

The happiness group includes sensory pleasure, pride in achievement, amusement, relief and contentment. My claim here is that these all share the same facial expression – some variation on the action of zygomatic major and the outer portion of orbicularis oculi (6+12 in FACS terms) – with only minor variations in the timing of the movements and their strength. Perhaps the voice may provide separate signals for each of these happiness emotions, for it certainly seems obvious that amusement doesn't sound like relief.

Before closing let me describe 'referential expressions', for these have been a source of confusion among those studying facial expression, and they are interesting in their own right. A referential expression is a facial movement that refers to an emotion that is not felt by the person showing it. The person is referring to the emotion, much as the person could with a word, showing the emotion but in a way that makes it clear that the person is not feeling it now.

The referential expression then must look like the emotional expression but differ from it for two reasons. First if it resembles the actual emotional expression the observer might fail to recognize that it is not being felt now, and perceive it as an actual emotional expression. The second reason why it must be transformed is to reduce the likelihood that the person making the referential expression will actually begin to experience an emotion. My research with Robert Levenson (Ekman *et al.* 1983; Levenson *et al.* 1990) found that deliberately making the facial movements associated with a universal expression generates the changes in the autonomic nervous system that occur when emotion is generated in more typical ways. In subsequent research we (Ekman and Davidson 1993) found that deliberately making the facial configuration associated with enjoyment generated many of the changes in the central nervous system which occur when enjoyment occurs spontaneously.

The transformations of the actual emotional expression into a referential expression occur in both time and morphology. Typically the referential expression is either much briefer or much longer than the actual emotional expression, and it involves only one set of facial movements, not the complete array of facial movements associated with an emotional expression. Common examples are the horizontal stretching of the lips, typically performed very quickly, to refer to fear, and the raised upper lip, either brief or very long, to refer to disgust.

A *caveat*: only what I have written about emotional expressions and emblems is based on systematic research in non Western as well as Western cultures. What little research that so far exists on illustrators, manipulators, and regulators has been done exclusively in Western cultures.

SUMMARY

I have offered a taxonomy of bodily and facial movement, taking account of communicative function, how the movement became part of the organism's repertoire, and the relationship between the movement itself and what it signifies. This five-way classification – emblems, illustrators, manipulators, regulators and emotional expressions – is based on semiotic, ethological and psychological perspectives on nonverbal signals. Although largely based on my first writing on this topic (Ekman & Friesen 1969), I have been able here to expand and reformulate my views based on findings obtained in the last thirty years.

Paul Ekman
University of California (San Francisco)
U.S.A.

REFERENCES

Banse, R. and Scherer, K.R. (1996). Acoustic profiles in vocal emotion expression. *Journal of Personality and Social Psychology*, **70**, 614–36.

Broide, N. (1977). Israeli emblems: a study of one category of communicate nonverbal behavior. Unpublished Master's Theses, University of Tel Aviv.

Darwin, C. (1998). *The expression of the emotions in man and animals. 3rd edition.* (ed. P. Ekman), HarperCollins, London and Oxford University Press, New York. (Original edition 1872, John Murray, London)

Dittman, A.T. (1972). Developmental factors in conversational behavior. *Journal of Communication*, **22**, 404–23.

Efron, D. (1968). *Gesture and environment.* King's Crown, New York.

Eibl-Eibesfeldt, I. (1970). *Ethology, the biology of behavior.* Hold, Rinehart and Winston, New York.

Ekman, P. (1972). Universals and cultural differences in facial expressions of emotion. *Nebraska symposium on motivation, 1971* (ed. J. Cole), pp. 207-83. University of Nebraska Press, Lincoln, Nebraska.

Ekman, P. (1979). About brows: emotional and conversational signals. In *Human ethology*, (ed. M. von Cranach, K. Foppa, W. Lepenies, and D. Ploog), pp. 169- 248. Cambridge University Press, Cambridge.

Ekman, P. (1992). An argument for basic emotions. *Cognition and Emotion*, **6**, 169–200.

Ekman, P. (1993). Facial expression of emotion. *American Psychologist*, **48**, 384–92.

Ekman, P. (1997). Expression or communication about emotion. *Uniting psychology and biology: integrative perspectives on human development* (ed. N. Segal, G.E. Weisfeld, and C.C. Weisfeld), pp. 315–38. APA, Washington, DC.

Ekman, P. (1999). Facial Expressions. *The handbook of cognition and emotion* (ed. T. Dalgleish and T. Power), John Wiley and Sons, Ltd., Sussex, U.K.

Ekman, P. (2001). Telling lies: clues to deceit in the marketplace, marriage, and politics 3th edition. W. W. Norton, New York.

Ekman, P., and Davidson, R. J. (1993). Voluntary smiling changes regional brain activity. *Psychological Science*, **4**, 342–45.

Ekman, P., and Friesen, W.V. (1969). The repertoire or nonverbal behavior: categories, origins, usage, and coding. *Semiotica*, **1**, 49–98.

Ekman, P. and Friesen, W. V. (1972). Hand movements. *Journal of Communication*, **22**, 353–74.

Ekman, P. and Friesen, W. V. (1974). Nonverbal behavior and psychopathology. *The psychology of depression: contemporary theory and research*. (ed. R. J. Friedman and M. N. Katz), pp. 203–32. J. Winston, Washington, D.C.

Ekman, P. and Friesen, W. V. (1978). *Facial action coding system: a technique for the measurement of facial movement*. Consulting Psychologists Press, Palo Alto, California.

Ekman, P., and Rosenberg, E.L. (ed.) (1997). What the face reveals: basic and applied studies of spontaneous expression using the facial action coding system (FACS). Oxford University Press, New York.

Ekman, P., Friesen, W. V., and Scherer, K. (1976). Body movement and voice pitch in deceptive interaction. *Semiotica*, **16** (1), 23–27.

Ekman, P., Friesen, W. V., O'Sullivan, M., and Scherer, K. (1980a). Relative importance of face, body, and speech in judgments of personality and affect. *Journal of Personality and Social Psychology*, **38**, 270–7.

Ekman, P., Roper, G., and Hager, J. C. (1980b). Deliberate facial movement. *Child Development*, **51**, 886–91.

Ekman, P., Levenson, R. W., and Friesen, W. V. (1983). Autonomic nervous system activity distinguishes between emotions. *Science*, **221**, 1208–10.

Johnson, H.G., Ekman, P., and Friesen, W.B. (1975). Communicative body movements: American emblems. *Semiotica*, **15**, 335–53.

Keltner, D. (1995). Signs of appeasement: evidence for the distinct displays of embarrassment, amusement, and shame. *Journal of Personality and Social Psychology*, **68**, 441–54.

Levenson, R. W., Ekman, P., and Friesen, W. V. (1990). Voluntary facial action generates emotion-specific autonomic nervous system activity. *Psychophysiology*, **27**, 363–84.

Schefflen, A.E. (1963). The significance of posture in communication systems *Psychiatry*, **26**, 316–31.

Schefflen, A.E. (1964). Communication and regulation in psychotherapy. *Psychiatry*, **27**, 126–36.

Schefflen, A.E. (1965). Quasi-courtship behavior in psychotherapy. *Psychiatry*, **28**, 245–57.

Trupin, C.M. (1976). Linguistics and gesture: an application of linguistic theory to the study of emblems. Unpublished doctoral dissertation, University of Michigan.

ISABEL GÓMEZ TXURRUKA[1]

FOCUS IN DISCOURSE: ALTERNATIVE SEMANTICS VS. A REPRESENTATIONAL APPROACH IN SDRT

1. INTRODUCTION

This paper compares two different theories of focus. On the one hand, alternative semantics, introduced in Rooth 1985, postulates a semantic interpretation of focus according to which the focus of a sentence is interpreted as a set of alternative propositions to the actual one. On the other hand, a representational approach to focus has been developed in SDRT (Asher 1995, Gómez Txurruka 1997) within which the focus/background structure introduces a partition of the representation contributed by the utterance. We first introduce both approaches and then attempt to use them to analyze the focus-discourse interface. Particular attention is paid to the interaction between focus and the discourse relation of Question-Answer Pair.

2. FOCUS IN ALTERNATIVE SEMANTICS

In Rooth 1985 focus is proposed to be interpreted as a set of alternative propositions that includes the sentence's ordinary meaning and at least one other different proposition. Rooth proposes a compositional way of constructing this interpretation. First, focus is assumed to introduce a feature F in the surface syntactic representation (in a Chomskian grammar). Second, the contribution of focus is computed semantically at LF in such a way that, besides the ordinary semantic meaning, we also get an alternative meaning. Consider, for example, the following sentence (capital letters signal an H* pitch accent in Pierrehumbert and Hirschberg's 1989 taxonomy; the slashes indicate the scope of the focus feature F):

Jesus M. Larrazabal and Luis A. Pérez Miranda (eds.), Language, Knowledge, and Representation: 51-68.

(1) Mary went to /PRAGUE/$_F$.

The alternative meaning, corresponding to the set of propositions of the form {||Mary went to Prague||, ||Mary went to Paris||, ...} is constructed by means of the following rules.

- Basic rules:

$\|\alpha\|_O = \|\alpha_F\|_O$

$\|\alpha_F\|_A = ALT(\|\alpha_F\|_O) = D_{Type(\|\alpha\|O)}$

$\|\alpha\|_A = \{\|\alpha\|_O\}$

If n is a proper name:

$\|n\|_O = \|n_F\|_O = \mathbf{n'} \in D_e$

$\|n_F\|_A = ALT(\mathbf{n'}) = D_e$

$\|n\|_A = \{\mathbf{n'}\}$

If V is an intransitive verb:

$\|V\|_O = \|V_F\|_O = \mathbf{V'} \in D_{<e,t>}$

$\|V_F\|_A = ALT(\mathbf{V'}) = D_{<e,t>}$

$\|V\|_A = \{\mathbf{V'}\}$

- Alternative compositional rule:

$\|\alpha\beta\|_A \approx \{X(Y) | X \in \|\alpha\|_A, Y \in \|\beta\|_A\}$

Using these rules we can compose the alternative meaning of the example above:

$\|Mary\|_A = \{\mathbf{mary'}\}$

$\|go\|_A = \{\mathbf{go'}\}$

$\|Prague_F\|_A = ALT(\mathbf{prague'}) = D_e$

$\|went\ to\ Prague_F\|_A = \{\mathbf{go'}(x) | x \in ALT(\mathbf{prague'})\}$

$\|Mary\ went\ to\ Prague_F\|_A = \{\mathbf{go'}(x, \mathbf{mary'}) | x \in D_e\}$

Thus, the alternative semantics does not separate Focus and Background at the representational level. The focus value is directly constructed in semantic terms: Focus contributes with a set of alternatives—namely, the set of propositions obtained (roughly) by substituting the focused lexical item with a variable that ranges over a domain D of the same type as the original item.

In Rooth 1992 some changes are introduced. In particular, it is proposed that the alternative set is not the whole set of propositions but just a subset of it:

(2) If C is the focus value, $C \subseteq ||\alpha||_A$.

(Moreover, C would include at least the ordinary meaning, $||\alpha||_O$, and one other element.) This restriction is intuitively correct as illustrated by (3), where the right set of alternatives to the focused element is not the set of propositions obtained by substituting in each object in the Domain but just those obtained by using the set {**prague'**, **paris'**}:

(3) A: Did Mary go to Prague or to Paris?

B: Mary went to $/\text{Prague}/_F$.

This example shows that the determination of the right subset in each case is highly context-dependent. In particular, information about the previous discourse seems to be crucial. Given an utterance, in order to obtain the relevant subset of alternative propositions, we need to restrict the variable of domain, D. D is already restricted to a particular semantic type, but this is insufficient to account for discourse coherence.

3. A REPRESENTATIONAL VIEW OF FOCUS IN SDRT

Let us now turn to how focus is handled in a representational way. This has been done in different frameworks such as DRT (see for example Kamp (1998, draft)) or the Structured Meaning Theory (Krifka 1991, 1996); here we will focus on the approach developed in SDRT (Asher 1995, Gómez Txurruka 1997).

SDRT (Asher 1993) is an extension of DRT. Suppose we have already built a representation of the first n-1 sentences in a discourse. In DRT, after processing sentence n, we just add the new conditions and discourse referents to the previous representation. SDRT replaces this union operation with a two-step procedure: (1) find which sites in the representation of the previous discourse (up to sentence n-1) are open (the representation furnished by sentence n-1 is always open) and (2) calculate which discourse relation might be used to attach the new representation to one of the attachment sites. In order to carry out these two steps, SDRT uses information provided by different sources (as lexical, semantic, or world knowledge, among others) and makes a strong use of an inference system that includes both a monotonic and a nonmonotonic entailment relation.

The basic idea to treat focus in a representational way is as follows. The focus of an utterance—or the so-called Focus/Background structure (F/B) of an utterance—introduces a partition of the representation of the utterance at some level prior to the semantic interpretation. In SDRT, this partition is carried out at the level of DRSs. Consider the sentence in (1) again (take the constituent between slashes to be the focus of the utterance):

(1) Mary went to /PRAGUE/.

Focus is inherited by the DRS-conditions. Let * be a Focus operator that applies to all the conditions introduced by the syntactic constituents in focus.[2] After processing (1), we get the following representation:

$x\ y\ e\ t_e$
Mary(x)
go-to(e,x,y)
*Prague(y)
$t_e < now$

We can directly obtain a partition of the set of conditions from this represetnation. If a condition is under the scope of the focus operator *, it is included in the set F; if not, it is included in B:

F={Prague(y)}

B={Mary(x), go-to(e,x,y), $t_e < now$}

Thus, note that, contrary to the interpretation provided by the alternative semantics, focus in SDRT is not taken to introduce directly a set; all it introduces is a partition.[3]

It has already been shown (see for example Asher 1995) that discourse information and discourse structure systematically interact with information partition. This interaction is bidirectional. On the one hand, different linguistic contexts can give rise to different information partitions for the same sentence with the same intonation pattern. Consider the following question-answer pairs:

(4) A: What did Mary do last year?

B: She /went to PRAGUE/.

C: Where did Mary go?

D: She went /to PRAGUE/.

If we want to distinguish direct answers to the question in (4A) from direct answers to (4C), different representations for (4B) and (4D) are needed. If (4B) is processed, the following DRS is obtained this time:[4]

$x\ y\ e\ t_e$
Mary(x)
*go-to(e,x,y)
*Prague(y)
t_e<now

From the DRS above, the following sets of conditions F and B are obtained:

F={Prague(y), go-to(e,x,y)}

B={Mary(x), t_e<now}

We can call the set of different information structures that a sentence with a given intonation pattern gives rise to the set of Potential Information Partitions (or set of PIPs).

The examples in (4) show that the same sentence with the same intonation has a different informational partition if it is an answer to the question in (4A) or to that in (4C). This means that different questions are related to different information partitions of the same answer with the same intonation. The same point can be made with other discourse structures such as Correction and Contrast. On the other hand, it is also easy to motivate the assumption that the Focus/Background structure plays a role in the determination of discourse structure and, in particular, that discourse relations such as Question-Answer Pairs, Corrections, and Contrasts require specific partitioning. For example, the correction in (5B) is not intuitively felicitous because the intonation is not appropriate in order for the listener to infer that a Correction is taking place as indicated by the discourse marker "no."

(5) A: Mary went to Prague.

B: #No, SHE went to Paris.

The particle "no" indicates Correction, but the presence of this discourse marker by itself is not enough for the listener to conclude that Correction holds; information partition also plays a role in the well-formedness of this relation. The same seems to obtain for discourse relations such as Contrast in (6):

(6) Bill lost the 400m LAST year. But he should win it THIS year.

The particle "but" indicates Contrast. Thus, the listener's task is to construct this discourse relation. Taking into account the informational content of both sentences, he (tentatively) infers that from the first clause (normally) follows some information that is in conflict with the content in the second clause (see for instance Knot and Mellish 1996). The information partition of both sentences seems to play a crucial role in this construction since focus marks the information that stands in opposition; intuitively we can paraphrase (6) as *Regarding Bill, there is a contrast between what happened last year and what will happen this year given that last year he lost the 400m and, thus, normally he should lose this year too, and the speaker is stating the contrary (its negation).*

How can we account for the interaction between discourse and information partition in SDRT? We suggest below that the sets of conditions furnished by the information partition have to be related to the partition of a previous representation. A function that takes into account the thematic roles of the arguments will take care of this job. The general scheme of the interaction between discourse and F/B structure in SDRT will be phrased in the following terms: conclude that certain discourse relation holds between the labels of two DRSs α and β, and that β has certain F/B structure, if and only if there is a mapping from the sets of conditions provided by β's F and B into some partition of α's set of conditions (or of some other DRS that is presupposed or inferred from α).

4. ALTERNATIVE MEANINGS IN DISCOURSE: TWO PROBLEMS

The interaction between discourse and focus in the alternative semantics has to be accounted for in terms of a (more or less sophisticated) possible worlds semantics, since the contribution of focus is directly constructed as a set of sets of possible worlds (each set of possible worlds supporting the truth of one of the alternatives). Remember that the basic idea in this framework is (1) to associate a focus value with the lexical items in the syntactic analysis in such a way that, if a leaf is not marked as focused, the focus value is the same as the ordinary semantic value, and if a leaf is marked as focused, the focus value is computed as a function of the ordinary value,[5] and (2) to use an alternative compositional rule to combine the basic values to construct the focus value of the sentence. This procedure does not permit us to use the information contributed by focus at a different level, prior to the level of interpretation (prior to the possible worlds semantics). This means that, if the alternative focus is going to account for the interaction between discourse and focus, this interaction has to be captured in terms of possible worlds and possible worlds sets.

Rooth 1996 proposes that the relation between the focus of a sentence and the previous discourse has to be construed in presuppositional terms (i.e., the focus value is defined as a presuppositional meaning, cf. page 279), stating that focus has to be "licensed" by the discourse. We present below two problems the alternative framework has to deal with. The first is related to the fact that focus does not introduce an existential presupposition. We suggest that, if this is so, the set of alternatives that constitutes the focus value cannot be related truth-conditionally to the previous discourse.

4.1 The set of alternatives is not truth-conditionally related to context

The focus value is defined as presuppositional in Rooth 1996 (p. 279). Following Stalnaker 1979, an utterance of ϕ presupposes a proposition p iff ϕ is felicitous in any given context c only when c entails p. However, since the focus value is a set a propositions (not a single proposition), the standard definition of presupposition cannot be directly applied.

We could try to transform the focus value into a single proposition. For example, if we could assume that focus carries an existential presupposition (in the sense that at least one of the alternatives has to be true in the context), we could form the disjunction of the alternatives. This new focus value would be a proposition and the standard definition of presupposition would apply.

However, it has already been shown (Vallduví 1990, Rooth 1996, 1999) that focus does not carry an existential presupposition. Consider Rooth 's 1996 example:

(7) A: Did anyone win the football pool this week?

B: I doubt it, because it's unlikely that MARY won it, and I know that nobody else did.

Rooth resorts to the projection test in Karttunen and Peters 1979. Karttunen and Peters show that the presuppositions of the complement of *It's unlikely that* are projected out of the scope of the adverb. For instance in (8):

(8) It's unlikely that John knew that Mary won the pool.

The presupposition triggered by the factive *to know* is projected out of the sentence. We can paraphrase (8) as *Mary won the pool and it's unlikely that John knew it*. Thus, if the focus in the clause *MARY won it* carried the existential presupposition that *someone won the football pool*, it should be projected out of the sentence. However, the speaker in (8) does not intend to communicate that *Someone won the football pool and it's unlikely that Mary did it and I know that nobody else did.* Assuming that focus carries an existential presupposition, and that therefore it is projected out of the scope of the adverb *unlikely,* we get an interpretation that is inconsistent and cannot correspond to what the speaker intends to communicate.

Vallduví 1990 makes the same point with a simpler example:

(9) A: My friends went to the beach on Sunday.

B: No, NOBODY went to the beach. It was too cold.

An utterance of the sentence in (9B) with focus on *nobody* will not intuitively be taken to presuppose that someone went to the beach. The discourse particle "no" is interpreted in the context as *A's friends did not go to the beach* and therefore, the immediate previous context would not support the existential presupposition that someone went to the beach.

To sum up: a semantics requiring an existential presupposition for focus is too strong. Rooth agrees that focus only introduces a set of alternatives and that "(t)his is weaker than an existential presupposition because such alternatives can be relevant without any of them necessarily being true." (Op. cit.: 291)

If focus does not carry an existential presupposition, the disjunction of the alternatives introduced by focus cannot be taken to be true in the context since it might be the case that all of them are false. This means that the focus value is a set of alternatives that is not truth-conditionally related to context.

How is this relation between the alternative focus and the previous discourse to be construed? In what sense can it be said that a context presupposes a set of alternatives, as in Rooth's 1996 definition, if their disjunction is not necessarily true in the context?

The other difficulty the alternative semantics has to deal with is a very naive formal problem that can probably be solved, but as it stands now it seems to us that the relation between the contribution of focus and the rest of the sentence in the focus value is reversed if the aim is to restrict the set of alternatives.

4.2 The relation between the contribution of focus and the rest of the sentence in the focus value is reversed if the aim is to restrict the set of alternatives

Consider the example in (10):

(10) A: Mary went to Prague (α).

B: No (β), Mary went to PARIS (δ).

Rooth 1996 uses the following "discourse grammar" to represent how the focus operator ~ (~C indicates that C is the focus value), the sentence with the relevant focus, and the previous discourse relate to each other:

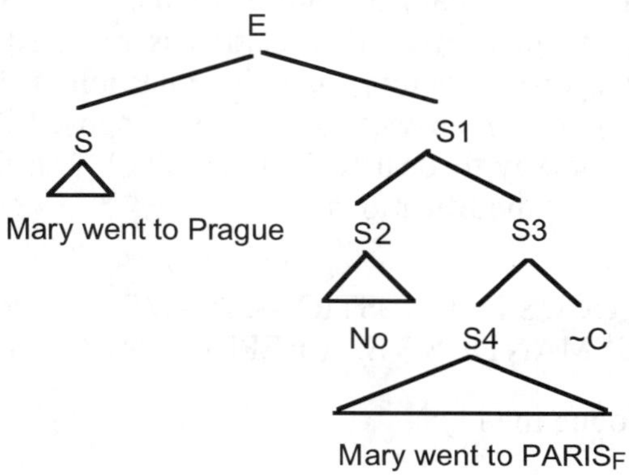

If C is the focus value, C is a subset of the set of alternatives ($C \subseteq ||\delta||_A$). The set of alternatives is as follows:

$||\delta||_A = \{||\text{Mary went to x}||: x \in D_e\}$

If one wanted to relate the focus value with a representational view of the Focus/Background partition, the "open proposition", ||Mary went to x||, would correlate with the Background, while the restriction of it, D_e, would correlate with Focus.

Intuitively, the focus value that the listener constructs taking into account discourse information would be the following pair set:

C={||Mary went to Prague||, ||Mary went to Paris||}.

Suppose, with Rooth, that we want to formally restrict the set of alternatives $||\delta||_A$ to obtain this particular subset C. Given the structure of $||\delta||_A$, to restrict $||\delta||_A$ we have to restrict D_e. As it stands, the domain comprises all the objects of type e. What information is needed to restrict D_e further?[6] We must use at least the information *Mary went to x*—which is information that in a representational view is given by the Background of the partition as well as being recoverable from the context. But in Rooth's possible worlds framework (even probably in a dynamicized possible worlds framework) this information is lost; it is irrecoverable either from the context or from the focus value. Consequently, D cannot be restricted.

Up to now, we have introduced two different approaches to focus, suggesting that the alternative account might face some problems. In particular, (1) since focus does not carry an existential presupposition, the alternative set is not truth-conditionally related to the context, and (2) we need the information in the "open proposition" to restrict the Domain of entities of the same type as the entity in focus, but in the focus value formula this relation is reversed. We turn next to a particular kind of discourse structure, namely, question-answer pairs. We first revise the approach to question-answer congruence proposed in Rooth 1996, and then propose a different way to deal with it in SDRT (we follow Asher 1995, introducing some relevant modifications).

5. FOCUS IN QUESTION-ANSWER PAIRS: ALTERNATIVE SEMANTICS VS. A REPRESENTATIONAL ACCOUNT

Consider the dialogue in (11):

(11) Q: Who went to Prague?

A: Mary went to Prague.

If questions are interpreted as sets of propositions, it seems a good move to also interpret the focus in the answer as a set, because it permits to account for the question-answer congruence by means of some relation between both sets. This is, in fact, the perspective taken in Rooth 1996.

There are at least two accounts of questions in terms of sets of answers. On the one hand, Groenendij and Stokhof 1984 propose that the meaning of a question is the set of exhaustive answers. That means that, if we take the domain in (11A) above to comprise just Mary and John, we get the following partition:

{||only Mary went to Prague||, ||only John went to Prague||,
||only Mary and John went to Prague||, ||Nobody went to Prague||}

Note, however, that this exhaustivity (from which it follows, for instance, that the first element of this partition entails that John did not go to Prague) is not part of the meaning of focus. Focus in *MAry went to Prague* does not convey that *ONLY Mary went to Prague*. The example in (12) supports this point:

(12) Q: Who went to Prague?

A: Well, MAry went to Prague; and also JOHN went to Prague.

Exhaustivity has been defined as a defeasible meaning that can be inferred from the Gricean maxim of quantity, but it does not seem to follow from the focus meaning. This means that the focus meaning is not going to match with any element of the partition. Groenendij and Stokhof's approach to questions does not seem appropriate to be combined with a theory of focus.

5.1 Question-Answer Congruence in Alternative Semantics

Rooth 1996 uses Hamblin's 1973 account. The meaning of a question is, with Hamblin, the set of possible answers. For example, the meaning of the question in (11) is as follows:

$||Q||= \{p| \exists x(\text{PERSON}(x) \wedge p = \mathbf{go'}(x, \mathbf{prague'}))\}$

In order to construct the interaction between a question and the focus in its answer, Rooth 1992 proposes that the meaning of the question should be included in the alternative set constructed from the focus in the answer: $||Q||\subseteq||A||_A$. Rooth 1996, on the other hand, suggests that the meaning of the question should be identical to the alternative set constructed from the focus in the answer: $||Q||=||A||_A$. Let us first consider how the constraint of inclusion works for the example in (11):

$$\{p| \exists x(\text{PERSON}(x) \wedge p=\mathbf{go'}(x, \mathbf{prague'}))\}\subseteq\{\mathbf{go'}(x, \mathbf{prague'})| x\in D_e\}$$

The relation of inclusion works fine in this example, given that the set of propositions where a person goes to Prague is included in the set of propositions where an entity of type e in the model goes to Prague (provided that the domain of entities e is not empty). Moreover, inclusion rules out utterances with inappropriate intonation such as:

(13) Q: Who went to Prague?

A: Mary went to PRAGUE.

The alternative set given by focus in (13A) is $\{\mathbf{go'}(\mathbf{mary'}, x)| x\in D_e\}$ and this is not a superset of $\{p| \exists x(\text{PERSON}(x) \wedge p=\mathbf{go'}(x, \mathbf{prague'}))\}$.

But Krifka 1998 shows that the relation of inclusion is not the right relation when multiple foci are involved. Consider the example in (14):

(14) Q. Who read Ulysses?

A_1. /MAry/$_F$ read Ulysses.

A_2. # /MAry/$_F$ read /ULYsses/$_F$ (and SUE read Moby-DICK).

Let us assume for the sake of the argument that the meaning of the question is as follows:

$\{p|\ p=||read(x, ulysses)||: x \in PERSON\}$

Let us further assume that this set is restricted in the context to the following:

{read'(mary', ulysses'), read'(susan', ulysses'), read'(john', ulysses')}

The alternative value of A_1 is given below:

$||A_1||_A=\{p|\ p = ||read(x, ulysses)||: x \in D_e\}$

The meaning of the question is a subset of the alternative set, that is:

{read'(mary', ulysses'), read'(susan', ulysses'), read'(john', ulysses')} $\subseteq ||A_1||_A=\{p|\ p = ||read(x, ulysses)||: x \in D_e\}$

Let us consider A_2 now. The alternative value of A_2 is the following:

$||A_2||_A=\{p|\ p = ||read(x, y)||: x \in D_e, y \in D_e\}$

The meaning of the question is also a subset of this set. The relation of inclusion wrongly predicts that the utterance in (14A_2) is felicitous in the context of (14Q).

As we said before, Rooth 1996 proposes identification. Note, however, that in (14), the meaning of the question is not identical to the alternative set of A_1 (the right relation being proper inclusion), that is,

$\{p|\ \exists x(PERSON(x) \wedge p=||x \text{ went to Prague}||)\} \neq \{||x \text{ went to Prague}||: x \in D_e\}$

This means that, if the aim is to exploit this requirement of identification to build the question-answer congruence, we must include a procedure that utilizes the information in the question to restrict the domain D_e in $||A_1||_F$. However, at this point, we come across the problems mentioned in §4 above. In particular (1) the non-focus part is not formally available if §4.2 holds; and (2) we cannot form the disjunction of the alternatives in order to ask that it be made true by the context if the argument in §4.1 is correct. Thus, if the arguments in §4 are right, this semantic procedure cannot make use of the non-focus part of the focus value and cannot treat the focus value as a single proposition in order to restrict the set of alternatives in the context. But then how can it be designed at all?

The semantic treatment of focus as a set of alternatives seems very attractive if questions are directly interpreted as sets of propositions. But questions can also be treated in a representational way, as we show below.[7] Furthermore, the interface between discourse and focus is not restricted to question-answer pairs. It must cover the discourse relations of Correction and Contrast too, as shown in Asher 1995 and Gómez Txurruka 1997. And note that in these cases it is not possible (or at least it does not seem the most direct route) to construct another set of alternatives from the information in the previous discourse.[8]

5.2 Focus and Question-Answer Pairs in SDRT

We next take a brief look at the interface between question-answer pairs and foci in Segmented DRT. We use mappings from the set of conditions obtained from processing the answer to the set of conditions obtained from processing the question. Let us first illustrate informally how these mappings work. The focus operator * produces a partition of the set of conditions in π_2. A function g operates on this partition, mapping both F(ocus) and B(ackground) to a partition of the set of conditions in π_1. The function g is defined in such a way that it takes into account the semantic structure of argument roles (or thematic roles). The picture below illustrates how this works for example (11):[9]

A: Who went to Prague? B: MARY went to Prague

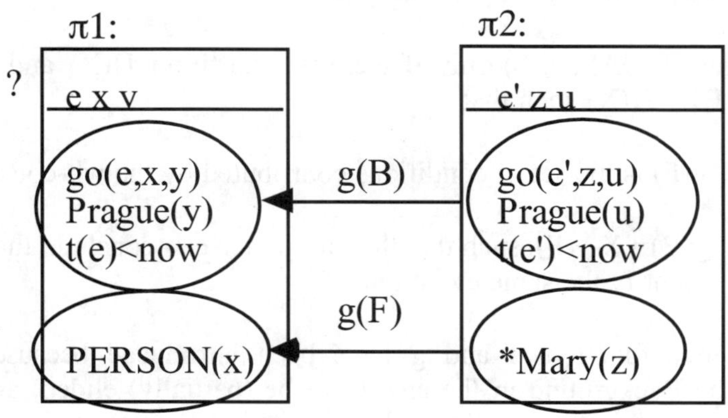

We restrict our attention to wh-questions including at most one wh-constituent and one event for simplicity. Answers will be direct and have an F/B structure. We refer to QAPs with these features as *simple wh-QAPs*.

Given DRSs α and β from a question and an answer respectively, let X and Y be their respective sets of conditions.

We assume that each condition is labeled with an argument role or with an event (we are leaving this information implicit).

We define first a Semantic Function f from the set of conditions Y to the set of conditions X, SemFun(f, Y, X):

$f: Y \rightarrow X$ such that $\forall b \in Y$,

(1) if b is labeled by an event condition, so is f(b), and

(2) if b is labeled by an argument role, f(b) is labeled by the same argument role (i.e., role(b)=role(f(b)).

SemFun(f, Y, X) can be used to define a function g from the partition of Y provided by F/B structure to some partition of X:

$g:\{F, B\} \rightarrow \Pi(X)$ is a function from {F, B} to a partition $\Pi(X)$ such that

$\exists f(\text{SemFunc}(f, Y, X) \wedge \text{if } \{b_1,...,b_n\} \in \{F,B\}, g(\{b_1,...,b_n\}) \supseteq \{f(b_1),...,f(b_n)\})$

We can define next a necessary condition on simple wh-QA congruence that will restrict the way in which the F/B structure of the answer relates to the question.

- **F/B Constraint on simple wh-QAP**:

  Simple wh-QAP(α, β) only if there is a partition $\Pi(X)$ and a bijection $g:\{F, B\} \rightarrow \Pi(X)$ such that

  (i) g(F) is the set of conditions contributed by the wh-constituent, and

  (ii) g(B)= X-g(F), such that the interpreter can conclude that both sets identify the same event type.[10]

Two different functions f and g have been introduced because it is not unusual for the Background in the answer to be (partially) elided, and we want the event introduced in the question to be inherited as Background in the answer.

On the other hand, condition (ii) will be able to handle differences in the way events and individuals are described, such as, for example, a change from an active to a passive structure, or use of different lexical items to refer to the same entity or property—what van Leusen's 1995 calls *perspective changes*.

CONCLUSION

We have first compared the different approaches to focus developed by the alternative semantics and SDRT. While the alternative semantics offers a semantic view, interpreting focus as a set of alternative propositions, focus in SDRT is taken to introduce a partition of a representation. We have pointed out, with Asher 1995, that focus interacts with discourse structure in Question-Answer Pairs, Corrections, and Contrasts, and therefore, that discourse structure has to be taken into account in any approach to focus.

We have next identified two possible problems alternative semantics has to deal with in order to be able to handle the interface between discourse and focus: (i) the alternative value is not truth-functionally related to context and (ii) the relation between the contribution of focus and the rest of the sentence is reversed if the aim is to restrict the focus value using discourse information.

Finally we have focused on an instance of the interaction between focus and discourse, namely Question-Answer Pairs. We have shown that the requirements of identification or inclusion proposed by Rooth 1992 and 1996 to account for the question-answer congruence are problematic, and an alternative way of constructing this interaction in SDRT has been offered.

Isabel Gómez Txurruka
ILCLI
The University of the Basque Country
Spain

NOTES

[1]I must thank the members of the University of Texas Tuesday Seminar on Discourse Semantics (and especially Ana Alves, Nicholas Asher, Cathy Berthouzouz, Tim Fernando and Yi Mao) for helpful comments and lots of motivation. The author gratefully acknowledges the sponsorship of the University of the Basque Country (UPV 003.230-HA 195/97) and the Department of Research of the Basque Country (post-doctoral grant 1997-98).

[2]In multiple focus cases, different focus operators will have to be introduced.

[3]It has been shown elsewhere (see for example Krifka 1991 or Asher 1995) that partitions are also appropriate to formally account for the association with focus. Focus sensitive operators such as *only* or *always* formally introduce a domain of quantification and the partition contributed by the focus/background structure helps restrict this domain.

[4]Note that we have not included the temporal condition (t_e<now) under the scope of *. The reason is that intuitively we want the time introduced by the event in (4B) to be identified with the time introduced by the event in the question. In order to perform this identification this condition should belong to the Background. This can be formally carried out if we use a syntactic representation that has an independent higher node to represent temporal information.
[5]See von Heusinger 1998 for details about how to calculate the focus value of determiners such as "the" and several problems.
[6]It is argued in Blok and Eberle 1999 that lexical and sortal information will further restrict this domain. Note, however, that this context-independent approach is not enough to carry out at least two tasks: (i) the selection of the appropriate information partition from the set of PIPs (see §3 above), and (ii) the construction of the actual Domain in the discourse (in (10B) above, D={**prague', paris'**}).
[7]See Krifka 1998 for a more developed account of questions in another representational framework, the structured meaning theory.
[8]Roberts 1996 proposes the systematic introduction of question meanings to construct the interface between focus and discourse. We have argued elsewhere that there is circularity in her definitions of focus and Question Under Discussion (see Gómez Txurruka 1998).
[9]We ignore the presupposition of existence usually associated to wh-questions.
[10]We could also use a non-pragmatic definition of individuation of events, as proposed in Carlson 1984, for example.

REFERENCES

Asher, Nicholas (1993): *Reference to Abstract Objects in Discourse*. Dordrecht: Kluwer Academic Press.

Asher, Nicholas (1995): 'From discourse macro-structure to micro-structure and back again: discourse semantics and the focus/background distinction.' *Context Dependence in the Analysis of Linguistic Meaning*, Volume I, Proceedings of the workshop in Prague (February/95) and Bad Teinach (May/95), Hans Kamp and Barbara Partee (eds.), Universität Stuttgart.

Blok, Peter I. and Kurt Eberle (1999): What is the alternative? The computation of focus alternatives from lexical and sortal information. In Peter Bosch and Rob van der Sandt (eds.), *Focus*. Cambridge: Cambridge University Press.

Büring, Daniel (1994): 'Topic,' in Bosch and van der Sandt (eds.), *Focus & Natural Language Processing*. Volume II: *Semantics*. Working papers of the Institute for Logic & Linguistics No. 7, Germany. (pp. 271-280).

Büring, Daniel (1997): 'The Great Scope Inversion Conspiracy,' *Linguistics and Philosophy* 20, 175-194.

Carlson, Greg N. (1994): 'Thematic roles and their role in semantic interpretation,' *Linguistics* 22, 259-279.

von Fintel, Kai (1994): Restrictions on quantifier domains. UMASS, Massachusetts: Ph. D. diss.

von Fintel, Kai (1995): A Minimal Theory of Adverbial Quantification. UMASS, Massachusetts: ms.

Gómez Txurruka, Isabel (1997): La Partición Informacional en el Discurso. University of the Basque Country, Spain: Ph. D. diss.

Gómez Txurruka, Isabel (1998): Are Foci Alternative Sets? ILCLI (University of the Basque Country), Donostia-San Sebastián: report No. ILCLI-98-LIC-5.

Groenendijk, Jeroen and Martin Stokhof (1985): *Studies on the semantics of questions and the pragmatics of answers.* University of Amsterdam: Ph. D. diss.

Hamblin, C. (1973): 'Questions in Montague Grammar,' *Foundations of Language*, 10, 41-53.

von Heusinger, Klaus (1998): Association with Focus in definite NPs. UCSC, Santa Cruz, CA: LRC-98-01.

Kamp, Hans (1998): A DRT-based treatment of the focus frame-focus division and its presuppositions: Some cases discussed in Rooth 1996. Universität Stuttgart: ms.

Karttunen, Lauri and Stanley Peters (1979): Conventional implicature, in Oh and Dinneen (eds.), *Syntax and Semantics*, 11. New York.

Knot, Alistair, and Chris Mellish (1996): 'A Feature-Based Account of the Relations Signalled by Sentence and Clause Connectives,' *Language and Speech* 39 (2-3), 143-183.

Krifka, Manfred (1991): A Compositional Semantics for Multiple Focus Constructions. University of Texas: second draft.

Krifka, Manfred (1996): Frameworks of Representation of Focus. ESSLI'96, Conference on Formal Grammar.

Krifka, Manfred (1998): Course on Pragmatics. Spring 1998. Austin: University of Texas.

Pierrehumbert, Janet and Julia Hirschberg (1989): 'The meaning of intonational contours in the interpretation of Discourse,' in Cohen, P.R., Jerry Morgan y Martha E. Pollack (eds.), *Intentions in Communication*. Cambridge, Ma: MIT Press, 271-312.

Roberts, Craige (1996): Information structure in Discourse: Towards and Integrated Formal Theory in Pragmatics. The Ohio State University: ms.

Rooth, Mats (1985): Association with focus. UMASS: Ph. D. diss.

Rooth, Mats (1992): 'A Theory of Focus Interpretation,' *Natural Language Semantics* 1, 75-116.

Rooth, Mats (1996): 'Focus,' in Shalom Lappin (ed.) (1996), *The handbook of Contemporary Semantic Theory*, Oxford, UK: Blackwell Publishers.

Rooth, Mats (1999): Association with focus or association with presupposition? In Peter Bosch and Rob van der Sandt (eds.), *Focus*. Cambridge: Cambridge University Press.

Stalnaker, R. (1979): 'Assertion', *Syntax and Semantics* 9.

Vallduví, Enric (1990): The Informational Component. University of Pen: Ph. D. diss. (Published in 1992, Garland Publishers.)

KEPA KORTA

ON COLLECTIVE INTENTIONS*

In those early amorphous years when memory had only just begun Esthappen and Rahel thought of themselves together as Me, and separately, individually, as We or Us.

Arundhati Roy: *The God of Small Things*, p. 2.

Unlike Esthappen and Rahel, I usually think of myself as Me, and I think as We or Us when thinking of myself together with other selves. I believe I am not a special case: most people (except, perhaps, real or fictional twins like Esthappen and Rahel) think of themselves in a similar way. When talking about Us, though, 'We' can be used in different occasions with different meanings. Leaving aside majestic uses and the like, sometimes 'We' merely refers to a set, a collection of people where 'I' is included. In this case, 'we' is nothing over and above the aggregation of individuals, an abbreviated manner of referring to her, you and me. But, sometimes we mean something stronger by 'We'. In this stronger sense, We is not yet a synonym for her, you and me; we somehow have been able to constitute together a new 'entity' called 'We'.

In this chapter, we will not discuss these different uses or meanings of the first-person plural pronoun. We will rather study the case of collective action; the sort of action whose agent is a 'we' in the second, strong sense; the sort of action that seems not to be explainable solely in terms of the actions of the individuals constituting such a 'we'. In fact, there is some consensus among philosophers on the view that there is such a thing as collective action, which is not just the sum of individual actions. However, there is no such a consensus on which its correct characterization is. Actions like singing a duet, dancing a tango, painting a house together, lifting the piano upstairs... are being intensely analyzed by philosophers like Raimo Tuomela, Margaret Gilbert, Michael Bratman and John Searle. The crucial concept is the concept of (*joint, shared, group, team, social, we-*) *collective* intention. The discussion of their proposals will be the main theme of this chapter.

Jesus M. Larrazabal and Luis A. Pérez Miranda (eds.), Language, Knowledge, and Representation: 69-91.

In the first section we will state as clear as possible the problem of collective action: how to reconcile the acceptance of the existence of collective action with the rejection of the existence of any primitive collective agent or mind. The second section will briefly stress the difference between collective action proper and a sum of singular actions. It is in the third section where we begin the discussion on the proposals mentioned above, with Gilbert's proposal and the risk to imply the existence of a group mind. Then, in the fourth section, which is the longest one, we will discuss Tuomela's, Bratman's and Searle's proposals, arguing for a primitive concept of collective intention. In the fifth section we will explain the place that this concept occupies in the explanation of collective action and agency. The sixth one will be devoted to the consideration of one important objection to any concept of collective intention. And the final section will be focused on some basic properties of collective intentions.

1. COLLECTIVE ACTION: THE PROBLEM

Let us assume from the very beginning, together with all the authors considered here, that there is such a thing as singular action. Each person can act alone in accordance with her own beliefs, desires and intentions, without taking into account, in principle, any other person's beliefs, desires and intentions. Of course, people usually share the physical environment so that they must coordinate their actions with the actions of other people: think in our behavior as drivers in the city traffic. But this does not make our actions collective; they are just an aggregate of singular actions, be they "social" or not in the Weberian sense. On the other hand, when we see a couple dancing a tango, two singers in a duet, a group of people painting a house together we speak naturally of collective action and collective agency,[1] meaning something else than a mere sum of individual actions. At an intuitive level it seems evident that

> "There really is such a thing as collective intentional behavior that is not the same as the summation of individual intentional behavior." Searle (1990), p. 402.

But philosophically this intuition turns out problematic. Any talk about the existence of something (action, mental state, agent) collective triggers, for some people, an immediate signal of danger. They surely think that the acceptance of anything collective necessarily undermines the so-called thesis of methodological individualism, which states that

> "All social phenomena ... should always be understood as resulting from the [mental states] and actions of individuals, and we should never be satisfied by an explanation in terms of so-called 'collectives'." Popper (1969), p. 98.

This is an ontological constraint. According to it, all social or collective behavior must be explained in terms of the mental states and actions of the individuals; in other words, any account of collective action should not involve in the ontology any other entity than individuals. Collective mental states ascribed to Hegelian collective minds are not to be accepted. Collective actions by mysterious superagents are not satisfactory explanations. All right. But accepting methodological individualism does not mean denying the existence of collective action. It only places a constraint, an important constraint, to our account of collective action. In fact, the authors that we will discuss here (Gilbert, Velleman, Searle, Tuomela, Bratman) all seem to accept this constraint. First, they accept the existence of collective action—this is what they want to explain. Second, they think that collective action is not just the mere aggregate of the actions of the individuals belonging to the relevant set of agents, not even when the agents share the same goal. But, third, they believe that, ontologically, a collective agent can be nothing over and above the individual agents that constitute it.[2] This means that the account of the existence of collective actions and agents should not be based on any postulation of collectives as ontological primitives. The only ontological primitives accepted are individuals. Quoting Searle again:

> "[Collective intentionality] must be consistent with the fact that society consists of nothing but individuals. Since society consists entirely of individuals, there cannot be a group mind or group consciousness. All consciousness is in individual minds, in individual brains." Searle (1990), p. 406.

But is it possible to make justice to our intuitions on collective action without violating this constraint? The proposals considered here try to show that it is.

2. COLLECTIVE ACTION VERSUS AGGREGATE ACTIONS

Being performed by more than one agent is, of course, a necessary condition for collective action. But it is not sufficient. As we said earlier, there can be a set of agents acting at the same time, even coordinately, without acting collectively. But what is more important, the same sort of bodily movements by a set of agents can count, in one case as a collective action, and as a mere aggregation of singular actions in another. Take a real example:[3]

In a holiday park where some hundreds of people are consuming drinks at the taverns especially opened for the occasion, a policeman appears in plain clothes and, recognized by some young guys, is beaten by them, and consequently seriously injured. In the action of beating the policeman each guy has the intention to beat him, but we can suppose, following the results of the judicial investigations well-known by now, that each guy's intention is in the relevant sense independent of the intentions and actions of the others. That was one of the main points to be established by the judge. He thought that in this case it was no collective action but a sequence of individual actions that happened to converge on a common goal.

So, the identification of the actions performed by a set of people either as a collective action proper or as aggregate singular actions cannot be determined solely by the bodily movements of the agents. We must consider the intentions of the agents. However, that an action is collective seems not to be fully explained by appealing to the singular actions and intentions of the agents involved, even when these actions have a common goal. We need a new type of intention, which we call collective intention. But what is this?

3. COLLECTIVE INTENTION AND GROUP MINDS

We want to characterize what a collective action is, i.e. in which consists an action performed by a collective agent. By analogy with the singular case, we can do it by ascribing to the agents a particular mental state called, *collective* (Searle), *shared* (Bratman, Gilbert) or *we-* (Tuomela) intention. All these terms may suggest, however, that it is not an intention of an individual, that it is not an intention of me or and intention of you, but an intention of us: something that, borrowing some graphical portraits from Searle (1995), might be represented like this:

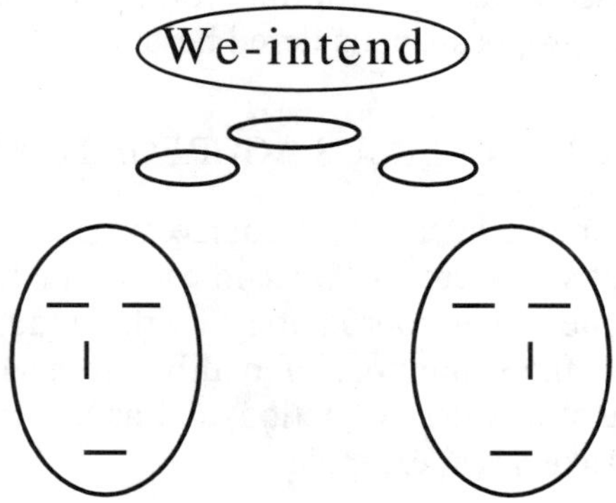

This sort of view, to my knowledge, is explicitly defended, in the recent philosophical literature on the matter, only by David Velleman:

> "... distinct intentions held by different people can add to a singletoken of intention, jointly held." Velleman (1997), p. 31.

I think that "a single token ofintention, jointly held" is the best description of the graphic above. Velleman wants to show how two or more individuals can literally share an intention. This aim immediately raises serious questions like where is this collective intention instantiated? In whose mind/brain? Not in mine, not in yours, but in ours? What does it mean? In answering these questions Velleman's proposal is—as, I think, any other along the same lines should be—irremediably led to the following dilemma: either (1) the term "intention" is no longer used for a mental state but for some other thing that can be literally shared, or (2) methodological individualism is abandoned and there is room for collective minds and consciousness. Velleman is ready to admit both horns of the dilemma. Concerning the first,

> "I am not sure that intention is essentially mental. There are of course mental intentions, but perhaps there can also be oral or written intentions—just as there are not only mental but also oral or written assertions." Velleman (1997), p. 37.

Concerning the second,

> "One may want to insist on intention's being a mental state, of course. But then I would be inclined to say that the existence of collective minds remains an open question." Velleman (1997), p. 38.

He then proposes to suspend judgment on both questions. But he is surely asking too much.[4] What we want is an account of collective action that, first and obviously, does not deny the existence of collective action, second, keeps the concept of intention as referring to a mental state within the mental cause theory of action and, third, takes methodological individualism as a serious constraint. Abandoning any of these principles seems just question begging.

Tuomela, Searle, Bratman, and Gilbert all say that they want their proposals to be consistent with methodological individualism. However, it is not always clear that, in fact, they are. In particular, Margaret Gilbert, with her account in terms of "plural subjects," seems often to imply the existence of a supermind. When she talks about shared intention, it is easy to understand that she is defending something similar to Velleman's concept of collective intention as a "a single token of intention, jointly held":

> "...it seems that there could be a shared intention to do such-and-such though none of the participants **personally** intend to conform their behavior to the shared intention" Gilbert (1997), p. 68.[5]

or, some lines below

> "... it is apparently possible in principle that the corresponding personal intentions be lacking when a shared intention is present." Gilbert (1997), p. 68-69.

It is worth trying to clarify what Gilbert is stating. I think we can read her words in two different ways. On the one hand, notice that she says **personal** intention, not **individual.** If we understand by "personal intention" just "singular intention" in the sense proposed earlier, i.efor the classical notion of intention considered traditionally in action theory, her claim would amount to saying that it is not necessary to have an intention of that kind once there is a shared or collective intention. If, according to her account, collective intentions were still mental states as individual as singular intentions, I would not pose any objection to those claims, and wait for her positive account of shared intentions. This interpretation seems to be consistent with her following words:

> "Does it mean that a shared intention is not a function of the mental states of the participants? **No**, only that the relevant mental states are not personal intentions of the participants in favor of the corresponding behavior. Does it mean that shared intentions cannot motivate? **No**, only that insofar as they do motivate, their motivational force does not derive from the motivational force of corresponding personal intentions. Does it mean that a sensitive understanding of the nature of *intention* is not relevant to an understanding of shared intention? **Not necessarily**." Gilbert (1997), p. 69.[6]

But these negative answers, on the other hand, do not sufficiently clarify her positions. Our doubts would have been removed had she answered "Yes, a shared intention is a (function of the) mental state(s) of the participants. Yes, shared intentions do motivate. Yes, a sensitive understanding of the nature of intention is relevant to an understanding of share intention."

But, as it stands, Gilbert's proposal can be taken to claim that shared intentions are not intentions of the individual minds, thus, implying either the existence of supraindividual minds or that shared intentions are not mental states.[7] And, as said before, we want a theory which does not imply the existence of any group mind and which takes collective intentions as mental states.

4. COLLECTIVE INTENTION: CAN IT BE REDUCED OR IS IT A PRIMITIVE CONCEPT?

Our account of collective action should explain how its existence is possible as different from the mere summation of singular actions, whereas we stick to methodological individualism, that is, to the ontological assumption that there are no primitive collectives, groups or 'we's. So the conceptual analysis goes in the following direction:

1. Collective intentions
   ⇓
2. Collective action
   ⇓
3. Collective agent, group, team, we...

('X⇒Y" is construed as 'X (or its analysis) plays an essential role in the analysis of Y but not vice versa.)

It is worth insisting that we assume that ontologically there should be no difference between a collective agent, a group or *we* and a mere set of agents. *We* are just you and I. But we are explaining the difference between, in the example above,

(a) 'We (all) hit the policeman,' that can be true even if you and me, each separately, hit him, with different singular intentions, and
(b) 'We hit the policeman (together),' that is not true in the case above, because *we* hit the policeman collectively. I hit the policeman and you hit the policeman but only as long as you and I are the constituents of this collective 'we'.

The intuitive difference between these two cases seems evident.[8] How do we explain it? Since we cannot appeal to any ontological difference between the 'we's in (a) and (b), we have to explain it recurring to the notion of 'collective action.' So a collective agent is just the term we use to refer to the (system of) agents of a collective action.[9] For clarifying the notion of collective agent, thus, we have to analyze the notion of collective action. It is important to note that now it will not do to define collective action as the action whose agent is collective.[10] That would make the analysis viciously circular. We have to define collective action recurring to the mental states of the agents that participate in it. This sort of mental states are what we call 'collective intentions', not because they are a mental state of a collective agent—this terminology is, of course, misleading, but we lack a better one—but because they define the actions of the agents as a collective action and, through this, the system of agents as a collective agent. We should keep in mind this direction of the conceptual analysis for approaching the main topic of the present section, namely, the discussion on the primitive or reductive analysis of collective intentions.

Once left aside those proposals that involve the existence of supraindividual minds, we can classify, following Searle (1990, 1995), the different explanations we are considering as belonging to one of these two groups:

Group A. Those who try to define collective intentions in terms of singular intentions and mutual beliefs.

Group B. Those who claim that collective intention is a primitive concept.

Of course, all the authors do not wholly agree on their ascriptions to one group or the other. According to Searle, the only author in B is himself, while Bratman and Tuomela (and Miller) are in Group A.

Bratman considers his own proposal "individualistic in spirit," thus including himself in group A, while in B he includes Searle and Gilbert:

> "In "Collective Intentions and Actions" John Searle argues that "collective intentional behavior is a primitive phenomenon" (401) and that we should eschew "a reductive analysis of collective intentionality" (406). In *On Social Facts* Margaret Gilbert focuses on "plural subject concepts" (...) In contrast with both Searle and Gilbert, I have argued that a useful reduction may be possible here." Bratman (1992), p. 341, fn. 25.

On the other hand, though sometimes his definitions seem to suggest the opposite, Tuomela has repeatedly claim that his proposal is not a reductive attempt:

> "we therefore need a concept of we-intention which is not reducible to mere personal I-intentions", Tuomela and Miller (1988).

> "Searle incorrectly assumes in his paper that (conceptual) reduction of we-intentions to personal intentions and mutual beliefs was attempted in our paper. (...) Searle (1990) adopts a view of collective intentions which is rather close to my view." Tuomela (1995), pp. 427-8, fn. 6.

So, it seems that, after all, Bratman is alone in group A.[11] But, is collective intention reducible to (definable in terms of) **singular** intentions? Searle (1990) thought that Tuomela and Miller's analysis of *we-intentions* was just an affirmative answer to this question. Their analysis is roughly the following one:

The agent A, who is a member of the collective G, has the collective intention (joint or we-intention, as they call it) to do X if and only if:

(i) A intends to do her/his part of X;
(ii) A believes that the opportunities of success of X obtain; especially that a sufficient number of the remaining members of the group (at least probably) will do their parts of X; and
(iii) A believes that the members of the group mutually believe that the conditions expressed in (ii) obtain.

The fact they use the phrase 'if and only if' and A's 'collective intention' on the left hand-side and simply 'intends' on the right hand-side is what makes their analysis appear to be a reductive attempt. If we analyze this as a reductive definition, so that A's intention (i) is viewed as a mere singular intention the definition cannot differentiate between a collective action proper and a mere sum of singular actions. Take a simple version of the example above, with only two people, you and I, hitting the policeman. In this case, we can have

X= we hit the policeman

my part of X= I hit the policeman

I can have my singular intention to hit the policeman, thus satisfying (i); I can believe that you will hit the policeman, so that X will obtain, thus satisfying (ii); I can believe that this is mutually believed between us, satisfying (iii), and still we can be in case (a), a summation of singular intentions plus mutual beliefs, and not in case (b), a proper collective action of *us* (in the relevant collective sense) hitting the policeman together.

So, if understood as a reductive attempt, Tuomela and Miller's analysis fails. That is what Searle (1990) showed. But, as already remarked, we should not take it as a reductive definition. We should not interpret intention (i) as a singular intention: when making reference to 'his part of X' it is implicitly involving *our* doing it *together:*

> "(...) it is a conceptual condition or presupposition of myintention to do my part of X that we indeed are about to participate in a joint action" Tuomela and Miller (1988), p. 373.

Clearly this does not serve as a reductive definition, since it is circular. Remember that collective intention is the first step of our conceptual ladder. We can include neither a collective action X nor a collective agent G in the analysans of its definition without circularity. Tuomela acknowledges that, and so I think that we should construe his proposal not as a reductive definition but as a reflexive, "fix-point," definition of a primitive concept of collective intention.

Bratman sees no vicious circularity in his own proposal. As he states in the quote above he makes his proposal as a reductive attempt. Then, he seems to belong to group A. But does he succeed in reducing collective intentions to singular ones? Let us examine his analysis.

Actually, it is not fair to represent Bratman's account by the graphic corresponding to group A. The intentions attributed to the agents participating in a collective action are, according to Bratman, not simply

I intend
but
(*) I intend that we do J,

where J is a collective action. On the face of it, this does not seem very different from a Searlean collective intention of the form 'We intend to J,' which is also an intention of an individual. According to Velleman's remarks, these different approaches would share the same strategy:

> "to imagine two or more agents as individually holding different token intentions of the same type, by holding intentions formulated in the first-person plural. The idea is that you and I can partake in the same intention if there is something that each of us individually intends that "we" are going to do." Velleman (1997), p. 33.

But there are, I think, important differences. First, Bratman's account

> "exploit the fact that we speak not only of intentions *to*, but also of intentions *that*—for example, my intention that Scott clean up his room. Accordingly, we can speak of my intention that we J." Bratman (1993), p. 102.

These 'intentions *that*' are not without problems. But let us assume that one can intend that we J.[12] He includes "we" and "J" (a collective action) in the definition of this intention. This can be circular. And Bratman acknowledges it:

> "Th e content of this intention[that we dance a tango together] seems already to bring in the idea of shared intentional activity. But I wanted to explain shared intentionality by appeal to the idea of shared intention. So I worried that we were threatened by an unacceptable circularity." Bratman (1997), pp. 51-52.

This is precisely one of my most important objections to Bratman's account: including a collective action J in the object of the collective intention he incurs in an unacceptable circularity. He proposes the following way out: According to him, in the case of singular action we avoid a similar circularity concerning intentionality by eschewing

> "appeal to attitudes that include in their content the very ideaof intentional action. Instead, we limit ourselves to contents that, at most, involve a concept of action that leaves it open whether the action is intentional." Bratman (1997), p. 52.

Similarly in the case of his 'intention that we J,' he proposes that

> "we try to understand J, as it appears in the content of my intention, ina way that leaves it open whether our J-ing is a case of shared intentionality." Bratman (1997), p. 52.

Notwithstanding, we should notice that there are different types of circularity involved here. In the case of singular action, he is talking about the so-called causal self-reference of intentions. Is it my intention to raise my arm, an intention to raise it intentionally? Are we not including the very idea of intentional action in the content of intention, when our strategy is to define intentional actions via intentions? As he said in the quote above, in this case we can leave it open whether the action is intentional.[13]

However, in the case of collective intentions there are two sorts of circularity involved: one coming from the intentionality of J, and the other from the sharedness, collective nature of J. The first is harmless, but what about the second? If we leave it open whether J is a collective action, how does our 'intention that we J' characterize J as a collective action? One possible answer would be by the inclusion of 'we' in the content of my intention, but that would be again clearly circular. When we consider Bratman's 'intend that we J' we should

> "assume that we will have available appropriate conceptions of joint activity that are neutral with respect to shared intention; or anyway, my discussion is limited to such cases." Bratman (1993), p. 101.

So, Bratman's intentions do not make any appeal to any previous idea of collective action or agency. What are, then, the attitudes of the agents that make their action collective? This is his proposal of a definition of shared intention:

> "We intend to J if and only if
>
> (a) I intend that we J and (b) you intend that we J
>
> I intend that we J in accordance with and because of 1a, 1b, and meshing subplans of 1a, 1b; you intend that we J in accordance with and because of 1a, 1b, and meshing subplans of 1a and 1b.
>
> 1 and 2 are common knowledge between us."
>
> Bratman (1993), p. 106.

This "interlocking web of intentions" together with the common knowledge about it is what constitutes Bratman's shared intention. Let us see whether it works. Remember that it has to distinguish between collective action proper and the sum (even if it is complex) of singular actions. Taking again our example, do the agents satisfy 1, 2, and 3? Action J will be 'hit the policeman,' with no collectivity built-in, and the same for 'we' in 'intend that we J'. (1a) is then equivalent to the conjunction of

> (1a') I intend that I J (which, I guess, means just 'I intend to J')
> and
> (1a'') I intend that you J.

So, if I intend to hit the policeman and I intend that you also hit him, I satisfy (1a), and similarly for you and (1b). It is not unnatural to think that, in the real case, the guys had just this type of intentions. They had each a singular intention to hit the policeman, and also an intention that the others do the same in order to get, for instance, the anonymity that provides acting within 'the mass.' And this can also explain that I intend that we J because (1a) and (1b); that is, if I intend that we J, then I intend that we J because I intend that we J (causal self-referentiality of intentions)[14], and also I intend that we J because you intend that we J. In other words, my intention that we J is dependent on your intention that we J, because I would not intend that we hit the policeman if I knew (or believed) that you did not intend that we hit the policeman.

We can also think that their intentions are in accordance and because of their "meshing subplans." This condition does not mean, following Bratman, that their subplans should completely match nor that they should be fully determined, but what is required is that "each of us *intends* that we J by way of meshing subplans" that guarantee appropriate interpersonal coordination.All this can be common knowledge among us. So our action would satisfy all conditions in Bratman's definition. Yet our action would not be collective in the required sense. It would not count as an action of a collective, but as the aggregation (though coordinated) of our singular actions.

Hence, Bratman's definition in his intended interpretation—i.e., without any idea of collectivity incorporated in the action J or the pronoun 'we'—does not give us a correct analysis of collective intention. Bratman's analysis falls short unless it incorporates an irreducible collective sense of 'we' or 'J.' But thus it would be a circular analysis, failing then to reduce collective intentions to singular ones.

So far, we have one good reason to take collective intention as a primitive concept, in the sense suggested by Searle and also by Tuomela. But what do we mean by "primitive"?

In the same way that we say that personal intention is a primitive concept because it is not reducible to beliefs and desires, we say that collective intention is primitive because it is not reducible to singular intentions and mutual beliefs. We are saying that among the primitive notions we have for designating mental states we should add 'collective intention' to the list of belief, singular intention, desire, goal, etc.[15] We mean that in our conceptual analysis

1. Collective intentions

$\Downarrow$

2. Collective action

$\Downarrow$

3. Collective agent, group, team, we...

we cannot take a previous step and analyze collective intentions in terms of other kind of singular mental states. It does not mean that we are accepting supraindividual minds and thus rejecting methodological individualism. Remember that in spite of their name collective intentions are *individual mental states*. So, how can we explain the existence of collective agents? We will consider it in the next section.

5. CONSTRUCTING US

The clearest account of the existence of collectives that is in accordance both with methodological individualism and the primitive nature of collective intentions is, in my opinion, Searle's (1995, 1997):

> "On my view, the existence of collective intentionality as a psychological primitive in the individual heads of individual agents does not commit one to a primitive ontology of actual human collectives. On the contrary, the basic ontology is that of individual human organisms and their mental states. The collective arises from the fact that collective intentionality is in the individual heads of individual organisms. The actual social collective consists entirely of individual agents with collective intentionality in their heads, nothing

> more. Ontologically speaking, collective intentionality gives rise to the collective, and not the other way around." Searle (1997), p. 449.

So we begin with collective intentionality, with collective intentions in our case, intentions that are characterized by an irreducible *we-ness*. It is having these intentions what makes possible the performance of a collective action by a set of individual agents, which constitute now a collective agent. So, this need not be considered as an ontological primitive:

> "My conclusion is that social collectives can be constituted by the fact that individual agents think of themselves as part of a collective without thereby supposing that the collective is an ontological primitive. The collective's existence consists entirely in the fact that there is a number of individual agents who think of themselves as part of the collective." Searle (1997), p. 450.

On the one hand, we have an explanation of the existence of collectives, which does not consider them ontologically primitive. On the other hand, we need a primitive notion of collective intention. This sort of intention is still as individual as any other mental state, and thus it is consistent with the hypothesis of the brain in the vat:

> "I could have all the collective intentionality that I could want in my head and still be radically mistaken. The fact that I have a we-intention does not by itself imply that other people share my we-intention, or even that there is a "we" that my intention refers to. I take myself to be engaging in collective behavior with other people, but whether or not I am in fact succeeding in engaging in collective behavior with other people is not a matter of the contents in my head. The existence of collective intentionality does not imply the existence of human collectives actually satisfying the content of that intentionality. But once you have collective intentionality, if it is *in fact* shared by other people, the result is more than yourself and other people: collectively you now form a group." Searle (1997), pp. 449-450.

A collective intention, then, is not an intention of a group, but an intention that individuals have if they form a group. This is another difference between our Searlean notion of collective intention or Tuomelian notion of we-intention and Bratman's shared intention:

> "Shared intention is an intention of the group. You and I together may have a shared intention to dance a tango. *I* cannot by myself have a shared intention to dance a tango, though I can intend to dance one with you" Bratman (1997), p. 50.

But what does it mean "intention of a group"? Might he be talking about supraindividual mental states, about group minds? I do not think so. He adds just immediately:

> "On my proposal I can also have an intention that *we* dance the tango; and it is such an intention on my part, together with an analogous intention on your part, which is central to *our* shared intention to dance the tango." Bratman (1997), p. 50.

The difference could be seen just as a terminological one. Searle's "collective intention" and Tuomela's "we-intention" stand for the intentions in each individual of the group, whereas "shared intention" describes a state of the group of agents when each of them has an intention "that we J." Bratman's term is more appropriate since for something to be *shared* or *collective* it seems that more than one agent is required. However, "shared intention" is still misleading because it is not really an intention but the conjunction and interrelation of the intentions of the agents in the group. Actually, the different terms point to a difference on perspective. Bratman is describing the group and the agents' mental states from an objective, God-eye like perspective. Searle, on the other hand, adopts a subjective perspective. The two perspectives are not incompatible, but it is important to make clear which one we are adopting for not viewing inexistent differences, and being able to detect the real ones.

By now, we have argued for a primitive notion of collective intention, i.e. non-reducible to any configuration of singular intentions and mutual beliefs, which stands for an *individual* mental state, and make justice to the idea of collective action as essentially different for singular action without violating the thesis of methodological individualism. The crucial question now seems to characterize collective intentions by their properties and by their relations with other mental states.[16] But before doing that, we should face an important objection to these analyses of collective intention, which alludes to an alleged constraint on the possible objects of intention.

6. THE OWNERSHIP AND CONTROL OF *OUR* ACTIONS

Traditionally, philosophers of action have considered that the object of one's intentions could only be one's own actions; one could only intend actions that are within one's power to do or bring about:

> "The proper objects of intending... seem limited to my actions (not the sun's) and to things I can do." (Baier (1970), p. 649.

That may seem reasonable, considering intentions as individual mental states that cause that one's bodily movements constitute an action. The cause of my action is my intention; the cause of your action is your intention. When I have a collective intention, that is, I intend and action of *ours,* am I violating this basic condition on intentions? If so, there could not be, by principle, any intention formulated in the first person plural, any collective intention.[17]

In fact, I think this can be a more serious objection when collective intentions are analyzed à la Bratman. As we have seen, his analysis involves intentions of the type "I intend that you J," and, therefore, taking as the object of one's intention the action of yours, would violate the ownership condition of the objects of intention. Indeed, within the mental cause theory of action would be bizarre to see my intention as the cause of your bodily movements. However, this kind of intentions is not necessary in the analysis of collective intentions. I suggest that one's collective intention has the same action-motivating power as one's singular intention. For instance, when I have the collective intention that we (you and me) push the car, I do not need any other intention, in particular any other singular intention to start my bodily movements. For us to push the car it is necessary that *I* push the car; in that sense my collective intention is enough for explaining my behavior. Do I need to intend that you also push the car and violate the above condition? No, what happens when I have a collective intention that we push the car is that I also believe that you have the corresponding collective intention. Or, further, I probably believe that is mutually believed between you and me that each has a collective intention to push the car. If I am right in my (mutual) beliefs, you will have your collective intention to push the car, and that will be the cause of your action: your intention, not mine.

The own action condition then, seems to pose no problem on collective intention, because when I have a collective intention I am not intending that *you* intend, I only believe (mutually with respect to you) that you already have a collective intention of the same type. Moreover, when we consider conditions in a Bratmanian sense as future-directed intentions, collective intentions have the same characteristics of singular intentions:

> "Our J-ing may serve as an end for my planning and my actions. I may constrain my plans so as not to settle on options incompatible, given my beliefs, with our J-ing. I may see myself as faced with a problem about how, given the end of our J-ing, this is to be achieved. Our J-ing may, that is, serve as an end in my plans, constrain those plans, and pose problems of means and preliminary steps for those problems." Bratman (1997), p. 53.

So, in these points collective intentions play the same characteristic role of singular intentions. And, as Bratman concludes, we are still without an argument showing that there is an intrinsic incoherence in this kind of intentions.

What are then the characteristic properties of collective intentions? We will devote the last section to the discussion of some of the most important ones.

7. SOME PROPERTIES OF COLLECTIVE INTENTION

For studying the properties of collective intention it can be convenient to use a formal notation. We will use the formula

$\mathrm{COLINT}_{xy}\ \alpha$

which can be read in a Bratmanian vein as

"I (x) intend that we (y and me, collectively) do α"

or in a Searlean way as

"We (y and me, collectively) intend to do α."

The important point is that they are intentions of the individual x, which subjectively characterizes α as a collective action, and x and y as a collective agent, as *we*. Now, from an objective point of view, for α to be a truly collective action, it must be true that α has been performed under the following conjunction:

$$\mathrm{COLINT}_{xy}\, \alpha \wedge \mathrm{COLINT}_{yx}\, \alpha.$$

When Bratman talks about the intention of a group, this seems to be what he is talking about. But now it is obvious that this is not an intention, not even a shared intention, but the conjunction of the collective intentions of the members of the group. Since what we want to clarify are the properties of this type of mental states and its relationship with other mental states of the agents, we adopt the subjective perspective.

In the last section, when talking about the beliefs associated to one's collective intention, we were pointing to one particular feature of collective intentions: their overt nature. Traditionally, singular intentions can be grouped into three classes in terms of the relationship between their satisfaction and their recognition by other agents: those which are neutral; those which are covert, i.e., intentions that are not to be recognized by other agents to be satisfied; and those which are overt, namely, whose recognition is necessary for their satisfaction. Communicative intentions are a typical case of the last class. Well, collective intentions are also overt.[18] Tuomela and Miller and also Bratman clearly are assuming that with the last clause in their respective analyses.

As we said before, when I have a collective intention to push the car I also believe that you have the corresponding collective intention. In our formal notation:

$$\mathrm{COLINT}_{xy}\, \alpha \rightarrow \mathrm{BEL}_{x}\, \mathrm{COLINT}_{yx}\, \alpha.$$

But, probably is not only a simple belief on my part concerning your collective intention. Due to the overt nature of collective intentions, I have to believe that you believe that I believe you have that corresponding collective intention. In other words I have the mutual belief (with respect to you) that you have that collective intention. Namely,

$$\mathrm{COLINT}_{xy}\, \alpha \rightarrow \mathrm{MB}_{xy}\, \mathrm{COLINT}_{yx}\, \alpha.$$ [19]

This captures the overt nature (from my subjective point of view) of your collective intention. But, surely my collective intention is also overt. So we would have:

$COLINT_{xy}\ \alpha \rightarrow MB_{xy}\ (COLINT_{xy}\ \alpha \wedge COLINT_{yx}\ \alpha)$.

This formula is intended to represent the fact that when I have a collective intention to do α with you, I assume that there is a reciprocal intention on your part, and that these intentions are overt between us. Of course, collective intentions are overt for the agents in the collective action, but with respect to other agents outside the group it could be either overt or covert, or neutral.[20]

It is important to note that my having a collective intention implies neither that you also have it nor that you believe I have it. In other words the formula above is consistent both with

$\neg COLINT_{yx}\ \alpha$

and

$\neg BEL_{y}\ (COLINT_{xy}\ \alpha)$.

So our concept of collective intention meets the condition of the brain in a vat.

We could go further on the description and analysis of the properties of collective intention with all the basic mental states of the agents involved in a collective action, and we could insert that analysis taking into account that intentions in general give rise to plans.[21] That, no doubt, would lead us to a logic of collective intention within a KD45 logic for beliefs, $KD4_{bg}5_{bg}$ for goals and $KD4_{bi}5_{bi}$ for intentions in the case of singular agency plus, specifically, a $KDD^{*}4_{bci}5_{bci}4_{mbci}4_{mbci'}$ for collective intentions.[22]

It seems evident that the clarification of so important a concept as that of collective intention can only be achieved by a deep interaction among a good conceptualization, a clear logical proposal of a calculus, and a rigorous semantics. This is the way to follow, in my view, and probably we are just at the beginning, in need of testing this framework in such different domains of application as systems of collective agency in Distributed Artificial Intelligence, pragmatic analysis of discourse, and collective agency in social psychology. That is the only way to avoid what seems to be the current greatest danger of "collectivizing" all psychological notions, mixing confusedly psychology with sociology and, by the way, not distinguishing what a mental state is and what is not.

Kepa Korta
ILCLI
The University of the Basque Country
Spain

* Earlier and substantially different versions of this paper were presented at the Sixth International Colloquium on Cognitive Science (ICCS-99) in Donostia, May 1999, and at the Seminar on Communication in Torino, October 1999. I wish to thank the audiences in both places for their helpful comments. This work was partially supported by a research grant of the University of the Basque Country (UPV I09.I09-HA010/99) and a research grant of the Spanish Ministry of Education (MEC HI1998-0051) for a joint project with the Centro di Scienza Cognitiva of the University of Torino. The major part of this version was completed during a stay at the Engineering and Science Lab in Harvard University. I wish to thank Barbara Grosz for making it possible. Thanks also to Jesus M. Larrazabal for his helpful comments on earlier drafts of the paper.

1 The terminology used by the different authors does not help clarifying the exact terms of the discussion. In the literature we find a variety of labels such us *collective, joint, shared, group, team, social, we-, common, mutual* for actions, intentions and beliefs, sometimes used as synonyms, sometimes in fancy combinations as *shared joint intention*, just for an instance. I will use *singular* versus *collective* actions, intentions, and beliefs. I prefer *singular* to *individual* or *personal*, because these when opposed to *collective* suggest that collective actions and attitudes are not actions and attitudes of the individuals, of the persons. In fact, *collective* is not without problems, since it seems to suggest that they are not attitudes of individual minds but of a Hegelian World Spirit, some supraindividual mind. The problem is precisely to explain collective action and agency without implying the existence of such enigmatic "supermind."

2 We are not addressing the case of structured social groups and institutions. Anyhow, the explanation of their existence is necessarily based on the explanation of the simplest cases of collective agency, which is the aim of our discussion. See Searle (1995), Tuomela (1995) or Gilbert (1989).

3 Bilbao, summer 1993.

4 See also Stoutland (1997), who argues for *social* attitudes, which are not mental states, ascribable to *social* groups, without arguing for a supraindividual mind. The price to be paid is to abandon the causal theory of action.

5 My emphasis.

6 Emphasis added.

7 This is reinforced by the examples she uses: On a walk, Tina and Lena have a shared intention to turn back in half an hour, but Lena intends to act contrary to the shared intention. (Gilbert (1997), p. 68). This is even stronger that the compatibility of shared intentions with lack of the corresponding personal (singular) intentions. According to Gilbert, here Lena can have simultaneously the shared intention to go on a walk with Tina and the personal (singular) intention not to go. As far as I know, no theory of intention and action makes room for contradictory intentions. Velleman also construes Gilbert's account of plural subjects as sounding "perilously close to talk of group minds or superagents." Velleman (1997), p. 30.

8 That can be seen clearer, when we think of the responsibility of each agent in each case.

9 We are restricting our discussion to collective action and intentions. As far as I know, initially this was particularly the subject of discussion. It is true that Searle

speaks about 'collective intentionality', which would include collective beliefs, goals, wishes, etc., but he lacks a detailed account of these notions. On the other hand, Gilbert (1987) and particularly Tuomela (1995, Balzer and Tuomela (1997), Miller and Tuomela (1999)) extend their analysis to collective beliefs, goals and attitudes in general. Concerning belief, it would be interesting to analyze what are the differences between collective belief, on the one hand, and the much better known concept of mutual belief, on the other.

[10] As proposed, for instance, by Hobbs (1990).

[11] Nevertheless, his point of departure seems to be the same as Searle's, i.e., the intuition that collective behavior is not just the sum of individual behavior, on the one hand, and methodological individualism, on the other:
"On the one hand, it is clearly not enough for a shared intention to paint the house together that each intends to paint the house... On the other hand, a shared intention is not an attitude in the mind of some superagent consisting literally of some fusion of the two agents. There is no single mind which is the fusion of your mind and mine." Bratman (1993), p. 98.

[12] See Bratman (1997) and Postema (1995) for some of these problems and plausible responses. Anyway, 'intentions that' weaken Bratman's claim that in his account
"the idea is not to introduce some fundamentally new and distinctive attitude. The attitude we are appealing to is intention" Bratman (1997), p. 102.
This claim has to be weakened, especially when he contrasts it with Searle's collective intention. Collective intentions are also intentions and, after all, they do not seem to be more primitive (new and distinctive) than intentions *that.*

[13] Or even we can just talk about bodily movements and represent intentions in the following way:

I intend (raise my arm, as a result of this intention)

This is circular. But the circularity is harmless. There is no infinite regress. Harman (1976, 1986) and Searle (1983), argue for the causal self-referentiality of intentions. In contrast, Bratman suspects that intentions are causally self-referential only in the case of the intentions in "shared intentional activity" or collective action (Bratman 1992).

[14]Bratman accepts that for this kind of intentions (see precedent note).

[15] As we said before, we limit ourselves to the case of collective intentions. See note 9.

[16] Miller and Tuomela (1999) propose a classification of goals with respect to their carriers and contents. I don't think this makes much sense when talking about full-blown collective intentions.

[17] This is the main subject addressed by Bratman (1997) and Postema (1995). See their arguments for allowing the possibility of collective intentions. Anyhow, Bratman's solution involves stipulating a distinction between *intentions to* and *intentions that*. I agree with Postema that "the plural perspective is not necessarily committed to impossible intentions." Postema (1995), p. 60. My arguments, however, do not rely on his.

[18] The difference between communicative intentions and collective intentions concerning their overt nature is clear: for the satisfaction of communicative intentions their recognition is necessary and sufficient; for collective intentions it is necessary but not sufficient.

[19] Where $MB_{xy}\, p \equiv BEL_x\, (p \wedge MB_{yx}\, p)$. This is the fixpoint version of one-sided mutual belief (Colombetti 1993). Again, we are interested in operators that represent mental states of the agents. $MB_{xy}\, p$ is a belief of x, and it is compatible with the nonexistence of the corresponding belief of y, that is, with $\neg BEL_y\, p$ and even with $\neg MB_{yx}\, p$.

[20] Related to the overt nature of collective intentions is what Tuomela calls the Collectivity Condition of satisfaction for collective intentions:

> "If my intention that we perform X together is satisfied, then, on quasi-conceptual grounds, also your intention that we perform X together is satisfied, and vice versa ... and each of us believes so." Tuomela (1998), p. 137.

[21] Future directed intentions function, according to Bratman, as input to practical reasoning. The adoption of an intention poses a planning problem for the agent. She has to adopt a plan to satisfy her prior intention. In these cases, the agent's collective intention will imply an intention of the following type:

$COLINT_{xy}\, \alpha \rightarrow INT_x\, \beta$, such that $\beta \in \alpha$ and the relation $\in$ is defined in this way: $\beta \in \alpha$ if and only if β enables α and $\beta \cup \{\delta_i\}$ generates α. (The set $\{\delta_i\}$ corresponds to the collection of actions that constitutes a plan to do α.)

In short, according to our analysis, x's having the collective intention to do α involves, the intention to do her part of the collective action α. This formulation captures, I think, both Tuomela and Miller's first clause of we-intentions and Bratman's reference to each participant's subplans of the joint action. I do not think that the agent has always to adopt a later intention of this kind. The collective intention to push the car can be the only intention who causes her pushing as part of our pushing, but there can be other actions more complicated and expanded over time that asks for this division of labor. Miller and Tuomela's (1999) classification of goals can be relevant but I do not pursue the matter here.

[22] See Korta and Larrazabal (2001). The axioms, in modal logic, are:

K. $COLINT_{xy}\, (\alpha \rightarrow \beta) \rightarrow (COLINT_{xy}\, \alpha \rightarrow COLINT_{xy}\, \beta)$

D. $COLINT_{xy}\, \alpha \rightarrow \neg COLINT_{xy}\, \neg\alpha$

D*. $COLINT_{xy}\, \alpha \rightarrow \neg INT_x\, \neg\alpha$

4_{bci}. $COLINT_{xy}\, \alpha \rightarrow BEL_x\, COLINT_{xy}\, \alpha$

5_{bci}. $\neg COLINT_{xy}\, \alpha \rightarrow BEL_x\, \neg COLINT_{xy}\, \alpha$

4_{mbci}. $COLINT_{xy}\, \alpha \rightarrow MB_{xy}\, COLINT_{xy}\, \alpha$

$4_{mbci'}$. $COLINT_{xy}\, \alpha \rightarrow MB_{xy}\, COLINT_{yx}\, \alpha$

REFERENCES

Baier, A. (1970), Act and Intent. *The Journal of Philosophy* 67, 648-658.

Balzer, W. and R. Tuomela (1997), A fixed point approach to collective attitudes. In G. Holmström-Hintikka and R. Tuomela (eds.). *Contemporary Action Theory. Vol. 2: Social Action.* Dordrecht: Kluwer, pp. 115-142.

Bratman, M. E. (1992), Shared Cooperative Activity, *The Philosophical Review* **101(2)**, 327-341.

Bratman, M. E. (1993), Shared Intention. *Ethics* **104**, 97-113

Bratman, M.E. (1997), I intend that *we J.* In G. Holmström-Hintikka and R. Tuomela (eds.). *Contemporary Action Theory. Vol. 2: Social Action.* Dordrecht: Kluwer, pp. 49-63.

Colombetti, M. (1993), Formal semantics for mutual belief, *Artificial Intelligence* **62**, 341-353.

Gilbert, M. (1987), Modelling Collective Belief. *Synthese* **73**, 185-204.

Gilbert, M. (1989), *On Social Facts.* London: Routledge.

Gilbert, M. (1997), What is it for us to intend? In G. Holmström-Hintikka and R. Tuomela (eds.). *Contemporary Action Theory. Vol. 2: Social Action.* Dordrecht: Kluwer, pp. 65-85

Harman, G. (1976), Practical reasoning. *Review of metaphysics* **29**, 431-63.

Harman, G. (1986), *Change in view.* Cambridge, MA: MIT Press.

Hobbs, J. R. (1990), Artificial Intelligence and Collective Intentionality: Comments on Searle and on Grosz and Sidner. In P. R. Cohen, J. Morgan and M. E. Pollack, *Intentions in Communication.* Cambridge, MA: The MIT Press, 1990, pp. 445-459.

Korta, K. and J.M. Larrazabal (2001), On the logic of collective intention. *The Bulletin of Symbolic Logic* 7(1), March 2001, 126-127.

Miller, K. and R. Tuomela (1999), What are collective goals? Draft.

Popper, K. (1969), *The Open Society and its Enemies.* London: Routledge.

Postema, G. (1995), Morality in the First Person Plural. *Law and Philosophy* **14**, 35-64.

Searle, J. (1983), *Intentionality: An essay in the philosophy of mind.* Cambridge: Cambridge University Press.

Searle, J. (1990), Collective intentions and actions. In P. R. Cohen, J. Morgan and M. E. Pollack, *Intentions in Communication.* Cambridge, MA: The MIT Press, 1990, pp. 401-415.

Searle, J. (1995), *The Construction of Social Reality.* London: Penguin.

Searle, J. (1997), Précis of *The Construction of Social Reality. Philosophy and Phenomenological Research* **57**, 427-458.

Stoutland, F. (1997), Why are Philosophers of Action so Anti-Social? In L. Alanen, S. Heinäma and T. Wallgren (eds.), *Commonality and Particularity in Ethics,* London: Macmillan, pp. 45-74.

Tuomela, R. (1995), *The Importance of Us: A Philosophical Study of Basic Social Notions.* Stanford: Stanford University Press.

Tuomela, R. (1998), Collective Goals and Cooperation. In Arrazola, X. et al. (1998), *Discourse, Interaction and Communication.* Dordrecht: Kluwer.

Tuomela, R. and K. Miller (1988), We-intentions, *Philosophical Studies* **53**, 367-389.

Velleman, J. D. (1997), How to share an intention. *Philosophy and phenomenological research* **57**, 29-50.

WILLIAM G. LYCAN

THE PLURALITY OF CONSCIOUSNESS[1]

My topics are consciousness. The plural is deliberate.

Both in philosophy and in psychology, «the problem of consciousness» is supposed to be very special. And the past fifteen years have seen an explosion of work on «it» by philosophers, psychologists, and neuroscientists as well. Besides the individual journal articles and books, there is now a vigorous *Journal of Consciousness Studies*; and John Benjamins Publishing Company's book series, *Advances in Consciousness Research*, is flourishing and includes some excellent works.

But if we look closely at some sample texts, we are struck by an astonishing diversity of topics that have gone under the heading of «consciousness».

1

Here is one sample text. Martin Davies and Glyn Humphreys begin the Introduction to their acclaimed anthology, *Consciousness* (1993), as follows.

> Consciousness is, perhaps, the aspect of our mental lives that is the most perplexing, for both psychologists and philosophers. The *Macmillan Dictionary of Psychology* contains as its entry for *consciousness*: 'the having of perceptions, thoughts, and feelings; awareness. The term is impossible to define except in terms that are unintelligible without a grasp of what consciousness means' (Sutherland, 1989, p. 90). [For whatever reason, Davies and Humphreys chose not to quote the entry's concluding sentence: 'Nothing worth reading has been written on it.'] On the side of philosophy, Daniel Dennett notes in *The Oxford Companion to the Mind* that consciousness 'is both the most obvious and the most mysterious feature of our minds' (Gregory, 1987, p. 160); and Thomas Nagel famously remarks (1974, p. 166), 'Without consciousness the mind-body problem would be much less interesting. With consciousness it seems hopeless.'
>
> These remarks might suggest that consciousness—indefinable and mysterious—falls outside the scope of rational enquiry, defying both scientific and philosophical investigation. But, in fact, the topic spans the history of psychology from William James until the present; and the past fifteen or twenty years have also seen an upsurge of philosophical interest in the place of consciousness in the natural order. [p. 1]

Jesus M. Larrazabal and Luis A. Pérez Miranda (eds.), Language, Knowledge, and Representation: 93-102.

These paragraphs assume, or seem to, that there is this one thing, consciousness (that is elusive and hard to define and perhaps outside the scope of rational enquiry). The authors later go on to qualify that idea, distinguishing some notions that are so different as to have little to do with each other. But there are still many readers who would be taken in by those opening paragraphs if they didn't read further.

And in fact, the anthologized essays are a grab-bag; some of them bear no discernible relation to others. The collection contains (besides philosophical articles on the distinct(!) topics of awareness, qualia, and subjectivity), «Theory and Measurement in the Study of Unconscious Processes,» «Impairments of Visual Awareness,» «Freud's Cognitive Psychology of Intention: The Case of Dora,» and «The Intentionality of Animal Action.» Brian Smith once said of computation that it «is a *site*, not a subject-matter»; «consciousness» is a flea market or jumble sale, not even a site.

2

By way of complaining about philosophers' habitual logic-chopping, distinction-mongering, verbal disputation, etc., my psychologist friend Paul Schulman once told me, «We don't care what anything is *called*; we just wanna *measure* it.» Damned right, too; what does it matter which problems have been called «consciousness»? Philosophers should not care what the distinct problems are called either. But everyone ought to care that they *are distinct*. Conflation of them is not merely a theoretical possibility. It has occurred, over and over again, resulting in countless confusions and errors. I urge us all to keep the distinct problems distinct. Not only will confusion and error thereby be avoided; I argue that if we divide we shall conquer.

Some of the topics and issues that have gone under the heading of «consciousness» are brutely empirical. Some are more abstractly theoretical. Some are outright philosophical and conceptual. I myself see these differences as differences of degree only, but they are still differences. One important difference, though far from the only one, is that of «how does» questions from «how possibly» questions —e.g., «How does a human subject/brain accomplish such-and-such a task?,» vs. «How could any theory of kind so-and-so possibly explain this refractory feature of human experience?» or «How could a mere information-processing system possibly have properties of this remarkable sort?»

Scientists sometimes pay lip service to one notorious «how possibly» question or another, and even voice that question in a tone of awed respect, but then proceed to announce that what we need is more information-processing models, or more neuroscience, ignoring the theoretical obstacle that drives the «how possibly» question, the obstacle that makes the relevant achievement seem impossible.[2]

For example, all too often we hear it suggested that advances in neuroscience will solve Thomas Nagel's (1974) and Frank Jackson's (1982) conceptual problem of «knowing what it's like»—or even that advances in neuroscience *have* solved it. This is grievously confused. For Nagel's and Jackson's claim is precisely that there is an irreducible kind of phenomenal knowledge that cannot be revealed by science of any kind. Nagel's and Jackson's respective «Knowledge Arguments» for this radical thesis are purely philosophical; they contain no premises that depend on scientific fact. Now, either the arguments are unsound or they are sound. If they are unsound, then so far as has been shown, there is no such irreducible knowledge, and neither science nor anything else is needed to produce it. But if the arguments are sound, they show that no amount of science could possibly help to produce the special phenomenal knowledge.[3] Either way, neither neuroscience nor any other science is pertinent.

And in general, no appeal to normal science is going to answer any philosopher's «how possibly» question, though it might happen that some hidden corner of normal science would lessen the force of such a question.

3

Here are some of the topics and issues that have gone under the «c-» heading.

- Consciousness as opposed to *un*consciousness (the «normal waking state»).
- Empirical questions of accessibility, attention, and reportability.
- The control of voluntary action.
- The temporal anomalies (color phi, the cutaneous rabbit, etc.).
- The Binding Problem(s).
- The development of the *self* concept.
- Some matters of intention and forming intentions.
- Putative «consciousness» deficits: The possession of information without awareness of that information.

    --Blindsight.
    --Semantic (etc.) priming.
    --Agnosias with «covert knowledge.»
    --Neglect, ditto.

So far, these topics are all empirical, or at least toward the empirical end of the spectrum. Some, like the temporal anomalies, are of more philosophical interest than others.[4] But now here are some more abstract and theoretical ones.

State/event consciousness: A state of a subject, or an event occurring within the subject, is a conscious as opposed to an unconscious or subconscious state or event iff the subject is aware of being in the state / hosting the event. In virtue of what is the subject aware of some of her/his own mental goings-on but unaware of others?

Introspection and introspective awareness, and privileged access to the internal character of one's experience itself. I am directly or «directly» aware of my own experience, or some of it, in a way that you cannot be; that needs explaining.

Qualia (strictly so called): The monadic, first-order qualitative features of apparent phenomenal objects. When you are experiencing a yellowy-orange after-image, what exactly is it that has the yellowy-orange color? For that matter, when you are (veridically) seeing a ripe banana, there is a corresponding yellow patch in your visual field. What ontological account is to be given of the yellowness of that patch (which might be as it is even if the banana were not real)? Bertrand Russell took it to be obvious, in need of no argument at all, that the bearers of such phenomenal color properties are nonphysical individuals, «sense data» as he called them; for impressive argument in addition, see Jackson (1977).

Homogeneity or grainlessness: The smooth continuous character of a phenomenal quality such as color, as contrasting with the discrete, particulate nature of the material of which we are made. The after-image is entirely yellowy-orange, yellowy-orange through and through, without gaps, and it has no part that is not yellowy-orange. But nothing in the physical world has no part that is not yellowy-orange. (Sellars, 1962, 1965, 1971.)

The intrinsic perspectivalness, point-of-view-iness, and/or first-personishness of experience, as discussed by Gunderson (1970), Nagel (1974) and others. In one way or another, our experience of our own mental states requires the adopting of a very special point of view; our experience of our external environment, though invariably from a point of view, is not perspectival in the same, deeper way.

Funny facts, or special phenomenal knowledge as allegedly revealed by the «Knowledge Arguments» mentioned above. It seems that the facts must be nonphysical facts.

The ineffability of «what it's like» (in the higher-order sense): One often cannot express in words what it is like to have a particular sensation. What is it like to experience the yellowy-orange of a yellowy-orange after-image? (It's *yellowy-orange*, yes, I heard you the first time; but can you tell someone in intrinsic, not comparative, terms, what it's like to experience visual yellowy-orange?)

The «explanatory gap» called to our attention by Joseph Levine (1983, 1993): Even if God were to assure us that, say, the Type-Identity Theory of mind is true and that such-and-such a conscious experience is strictly identical with a firing of certain neural fibers, we would still lack an explanation of why those fiber firings feel to their subjects in the distinctive way they do. Indeed, to Levine it seems «arbitrary» that they do.

4

Psychologists and (especially) philosophers have tended to think of «consciousness» in the same mental breath as «phenomenal experience.» But notice that phenomenal experience is necessarily involved only in the last six of the foregoing issues, and has nothing intrinsically to do with any of the others. (I shall call the last six, qualia-strictly-so-called through the explanatory gap, «the problems of phenomenal experience.») Just those six are perhaps a daunting array.

Theories «of consciousness» have been offered by cognitive psychologists (Mandler (1985), Baars (1988), Shallice (1988a, 1988b), Johnson-Laird (1988), Schacter (1989), Kosslyn and Koenig (1992)), and by neuroscientists (Edelman (1989, 1992); Crick and Koch (1990)). Philosophers have put forward others: D.M. Armstrong's (1968, 1981) Lockean «inner sense» theory, David Rosenthal's (1993) Higher-Order Thought theory, D.C. Dennett's (1991) Multiple Drafts theory. However, the key thing to grasp about all of these is that not one of them even *addresses* any of the six problems of phenomenal experience. I am not speaking pejoratively and I do not mean my previous assertion as a criticism of any of the theories. Some of their creators did not really aim them at any of the six problems, even though they paid lip service as recorded above. Others, particularly the philosophers, did not even superficially intend them to address the six problems. The theories were aimed at different phenomena and may be admirable explanations of those phenomena.

I draw two morals: First, no one should claim that problems of phenomenal experience have been solved by any purely cognitive or neuroscientific theory. (I find myself in surprising agreement with Chalmers (1996) on that.) Second and perhaps more importantly, the theories cannot fairly be criticized for failing to illuminate problems of phenomenal experience. And many of them have been so criticized, e.g., by Chalmers (1996).[5] Armstrong's and Rosenthal's theories, in particular, are very explicitly theories of *awareness* and of privileged access, not theories of qualia or of subjectivity or of «what it's like.» (In fact, both Armstrong and Rosenthal offer theories of qualia, but quite different theories, and elsewhere (Armstrong (1968, 1999), Rosenthal (1991)). Never criticize a philosophical theory for not explaining X unless the theory is either a theory of

X or a theory of something which essentially includes X —at least not when its proponent quite rationally disavows concern for X.

5

It is the problems of phenomenal experience that have most greatly exercised philosophers, often under the heading «'the' problem of phenomenal experience.» But even once we have split off the six from the preceding ten and more, we must continue to distinguish and divide. Much harm has resulted, within philosophy, from conflating among the six themselves. To make progress, we must take them one at a time.

For the record —well, not only for it— I will note my own positions on each of the six problems,[6] and then conclude with a remark on why the positions are dialectically important.

Qualia (strictly so called): I am what has come to be called a Representationalist about the qualitative features of apparent phenomenal objects. When you see a (real) ripe banana and there is a corresponding yellow patch in your visual field, the yellowness «of» the patch is, like the banana itself, a representatum, an intentional object of the experience. The experience represents the banana and it represents the yellowness *of the banana*, and the latter yellowness is all the yellowness that is involved; there is no mental patch that is itself yellow. If you were only hallucinating a banana, the unreal banana would still be a representatum, but now an intentional inexistent; and so would be its yellowness. The yellowness would be as it is even though the banana were not real. Likewise, when you experience a yellowy-orange after-image, you visually represent a colored spot in real physical space, and the yellowy-orange is the represented spot's represented color.

Homogeneity or grainlessness: Here I would appeal to the Representational theory of qualia, and point out that vision does not represent smoothly colored physical objects as having gaps in color. (It is a further and interesting question whether this is tantamount to vision's representing the objects as having no gaps in color.)

The intrinsic perspectivalness, point-of-view-iness, and/or first-personishness of experience: I defend an «inner sense» theory of awareness and introspection that I swiped from Armstrong (1968, 1981). On this view, introspection is much like internal perception, and (like any mode of perception) gives you a unique and partial perspective on what you are perceiving. Indeed, I hold that introspection is the operation of an internal scanner or monitor, that produces representations of your first-order mental states themselves, representations that are made of concepts peculiar to the monitoring device. The blind men and the elephant are representing the same thing very differently, so differently that they do not, perhaps cannot realize that what they are representing is the same as what each of the other men are representing. Similarly, seeing an event and hearing the same event may be nothing alike. And introspecting what is in fact a neural event is not a bit like seeing that same event as from outside one's head, using mirrors. Your introspective device deploys very distinctive concepts of its own. These features of the introspector combine, I hold, to explain the intrinsic perspectivalness of the mental as such: The mental as such is the neural, not as such, but as viewed from the unique introspective perspective.

Funny facts and special phenomenal knowledge: I believe that there is special phenomenal knowledge. Jackson's (1982) character Mary the color-blind color scientist knows all the objective, physical facts about color and color perception, but when cured of her malady she herself experiences colors for the first time, and learns a new fact, the fact of what it is (actually) like to see red. On my view, the sense in which this is a new fact is that fine-grained sense according to which «facts» are not merely chunks of spacetime, but incorporate modes of presentation, or concepts under which the relevant chunks of spacetime are represented: thus, in this sense, that water is splashing and that H_2O molecules are moving in such-and-such a way are different «facts,» though in a more obvious coarse-grained sense they are one and the same fact. In the fine-grained sense, that I weigh 200 pounds and that WGL does are different facts; that the meeting begins at noon and that it begins five minutes from now are different facts; etc. In none of these cases is there any threat to the materialist view of human beings, but only different perspectives on the same chunk of spacetime. So too, when Mary learns that it is like…*this* [however she might classify the experience to herself] to see red, she is representing the qualitative character of the experience from the introspective perspective using an introspective concept, thereby generating a new finely-individuated fact, even though the spacetime chunk she is representing from that highly distinctive point of view is itself scientifically and naturalistically unremarkable.

The ineffability of «what it's like»: Ditto. When Mary represents the qualitative character of the experience of red, or yellowy-orange, from the introspective perspective she mobilizes an introspective concept, one that is proprietary to her introspector and (for a reason I cannot go into here) does not translate into English or any other natural language. Of course, the *coarse-grained* fact that is being reported by her introspector can be expressed and described in many ways, corresponding to many different fine-grained «facts»; it is only the fine-grained fact incorporating the special introspective concept that has no natural-language translation.

The «explanatory gap»: I agree that the Gap is real. But this is for two reasons, neither of which embarrasses materialism. First, as I have said, phenomenal information and facts of «what it's like» are ineffable. But one cannot explain what one cannot express in the first place. (The existence of ineffable facts is no embarrassment to science or to materialism, so long as they are fine-grained «facts,» incorporating modes of presentation. It is the modes that make them ineffable, not the underlying coarse-grained fact.) Second, the Gap is not confined to consciousness in any sense or even to mind; there are many kinds of intrinsically perspectival (fine-grained) facts that cannot be explained. Pronominal modes of presentation again serve as a good example. Suppose an opthalmologist explains why WGL is nearsighted. That does nothing to explain why I am nearsighted; nor could anyone or anything explain that—unless, of course, one *first* conceded the identity of me with WGL.

6

Of course I do not expect anyone (yet) to be convinced of my solutions to the six problems of phenomenal experience, when all I have offered are cryptic summaries of them and without a hint of argument. (I do hope you will be moved to go out and buy my books, preferably in hardcover.) But my solutions *are solutions*, in the sense that if correct they would solve the respective problems. I believe each one is correct. I may be wrong in one or more cases or even in all. But remember that the most difficult problems of phenomenal experience are expressed as «how possibly» questions («How could a purely physical organism be directly acquainted with qualia?»; «How could materialism permit the existence of intrinsically perspectival, ineffable facts?»). A «*here's* how possibly» answer must be treated with respect unless and until it is refuted. Of course, the solutions must be mutually consistent, as (so far as I can see) mine are. But I hope I have gone some way toward persuading you that each of the problems becomes more tractable once it has been carefully distinguished from the others.

William G. Lycan
Chapel Hill, NC. (U.S.A.)

NOTES

1 This talk was originally entitled «Cognitive Theories of Consciousness.» Thanks to Martin Davies and to Manuel Liz for their excellent comments on it at the ICCS_99 conference.

2 This theme has also been emphasized, and well documented, by Block (1995) and by Chalmers (1996).

Baars (1995) rejoins against Block that «[w]hether consciousness is a 'mongrel problem' or purebred is an empirical question» (p. 249). He gives the example of cancer, which for a time was considered a false natural kind but, he says, is now again regarded as «the same underlying dysfunction, expressed by different pathways and in different tissues» (p. 249). True or not, this squarely misses the point: Different senses of «consciousness» may be found empirically to have the same referent, but that would be—just that—an empirical discovery, not something to be simply assumed or even to be established by a priori argument.

3 For the record, I myself believe that the Knowledge Arguments' conclusion is equivocal. On one reading, it is entirely unproven and the arguments are simply unsound; on the other, it is proven and true, but not terribly interesting and in particular no threat to materialist theories of mind. See Lycan (1996), Chs. 3 and 5.
4 For a wonderful discussion of the temporal anomalies, see Chs. 5 and 6 of Dennett (1991).
5 See also Block (1993) and Goldman (1993).
6 Lycan (1987), (1996). I will not rehearse any of the arguments for the positions, but you may take my word for it that they are decisive.

REFERENCES

Armstrong, D.M. (1968). *A Materialist Theory of the Mind*. (London: Routledge and Kegan Paul.)

Armstrong, D.M. (1981). «What is Consciousness?» In *The Nature of Mind*. (Ithaca, NY: Cornell University Press.)

Armstrong, D.M. (1999) *The Mind-Body Problem: An Opinionated Introduction*. (Boulder, CO: Westview Press.)

Baars, B.J. (1988). *A Cognitive Theory of Consciousness*. (Cambridge: Cambridge University Press.)

Baars, B.J. (1995). «Evidence that Phenomenal Consciousness is the Same as Access Consciousness,» *Behavioral and Brain Sciences* 18: 249.

Block, N.J. (1993) Review of D.C. Dennett, *Consciousness Explained*. *Journal of Philosophy* 90: 181-93.

Block, N.J. (1995). «On a Confusion about a Function of Consciousness.» *Behavioral and Brain Sciences* 18: 227-47.

Chalmers, D. (1996). *The Conscious Mind*. (Oxford: Oxford University Press.)

Crick, F., and C. Koch (1990). «Towards a Neurobiological Theory of Consciousness.» *Seminars in the Neurosciences* 2: 263-75.

Davies, M., and G. Humphreys, eds. (1993). *Consciousness*. (Oxford: Basil Blackwell.)

Dennett, D.C. (1991). *Consciousness Explained*. (Boston: Little, Brown & Co.)

Edelman, G. (1989). *The Remembered Present: A Biological Theory of Consciousness*. (New York: Basic Books.)

Edelman, G. (1992). *Bright Air, Brilliant Fire*. (New York: Basic Books.)

Goldman, A.I. (1993). «Consciousness, Folk Psychology, and Cognitive Science.» *Consciousness and Cognition* 2: 364-82.

Gregory, R.L., ed. (1987). *The Oxford Companion to the Mind*. (Oxford: Oxford University Press.)

Gunderson, K. (1970). «Asymmetries and Mind-Body Perplexities.» In M. Radner and S. Winokur (eds.), *Minnesota Studies in the Philosophy of Science, Vol. IV*. (Minneapolis: University of Minnesota Press.)

Jackson, F. (1977). *Perception*. (Cambridge: Cambridge University Press.)

Jackson, F. (1982). «Epiphenomenal Qualia.» *Philosophical Quarterly* 32: 127-136.

Johnson-Laird, P. (1988). «A Computational Analysis of Consciousness.» In Marcel and Bisiach (1988).

Kosslyn, S., and O. Koenig (1992). *Wet Mind: The New Cognitive Neuroscience*. (Glencoe, IL: Free Press.)

Levine, J. (1983). «Materialism and Qualia: The Explanatory Gap.» *Pacific Philosophical Quarterly* 64: 354-61.

Levine, J. (1993). «On Leaving Out What It's Like.» In Davies and Humphreys (1993).

Lycan, W.G. (1987). *Consciousness*. (Cambridge, MA: Bradford Books / MIT Press.)

Lycan, W.G. (1996). *Consciousness and Experience*. (Cambridge, MA: Bradford Books / MIT Press.)

Mandler, G. (1985). *Cognitive Psychology*. (Hillsdale, NJ: Lawrence Erlbaum.)

Marcel, A.J., and E. Bisiach (eds.) (1988). *Consciousness in Contemporary Science*. (Oxford: Oxford University Press.)

Nagel, T. (1974). «What Is It Like to Be a Bat?,» *Philosophical Review* 83: 435-50.

Rosenthal, D. (1991). «The Independence of Consciousness and Sensory Quality.» In E. Villanueva (ed.), *Philosophical Issues, I: Consciousness*. (Atascadero, CA: Ridgeview Publishing.)

Rosenthal, D. (1993). «Thinking that One Thinks.» In Davies and Humphreys (1993).

Schacter, D.L. (1989). «On the Relation Between Memory and Consciousness: Dissociable Interactions and Conscious Experience.» In H. Roediger and F. Craik (eds.), *Varieties of Memory and Consciousness: Essays in Honour of Endel Tulving*. (Hillsdale, NJ: Lawrence Erlbaum.)

Sellars, W. (1962). «Philosophy and the Scientific Image of Man.» In R. Colodny (ed.), *Frontiers of Science and Philosophy*. (Pittsburgh: University of Pittsburgh Press.)

Sellars, W. (1965). «The Identity Approach to the Mind-Body Problem,» *Review of Metaphysics 18*: 430-51.

Sellars, W. (1971). «Science, Sense Impressions, and Sensa: A Reply to Cornman,» *Review of Metaphysics* 24: 391-447.

Shallice, T. (1988a). *From Neuropsychology to Mental Structure*. (Cambridge: Cambridge University Press.)

Shallice, T. (1988b). «Information-Processing Models of Consciousness: Possibilities and Problems.» In Marcel and Bisiach (1988).

Sutherland, N.S. (ed.) (1989). *Macmillan Dictionary of Psychology*. (London: Macmillan).

TOM M. MITCHELL

THE ROLE OF UNLABELED DATA IN SUPERVISED LEARNING

1. INTRODUCTION

Most computational models of supervised learning rely only on labeled training examples, and ignore the possible role of unlabeled data. This is true both for cognitive science models of learning such as SOAR [Newell 1990] and ACT-R [Anderson, et al. 1995], and for machine learning and data mining algorithms such as decision tree learning and inductive logic programming (see, e.g., [Mitchell 1997]). In this paper we consider the potential role of *unlabeled* data in supervised learning. We present an algorithm and experimental results demonstrating that unlabeled data can significantly improve learning accuracy in certain practical problems. We then identify the abstract problem structure that enables the algorithm to successfully utilize this unlabeled data, and prove that unlabeled data will boost learning accuracy for problems in this class. The problem class we identify includes problems where the features describing the examples are redundantly sufficient for classifying the example; a notion we make precise in this paper. This problem class includes many natural learning problems faced by humans, such as learning a semantic lexicon over noun phrases in natural language, and learning to recognize objects from multiple sensor inputs. We argue that models of human and animal learning should consider more strongly the potential role of unlabeled data, and that many natural learning problems fit the class we identify.

2. A SUPERVISED LEARNING PROBLEM: LEARNING TO CLASSIFY WEB PAGES

To illustrate the role of unlabeled data in supervised learning, consider the problem of learning to classify pages of hypertext from the world wide web, given labeled training data consisting of individual web pages along with their

Jesus M. Larrazabal and Luis A. Pérez Miranda (eds.), Language, Knowledge, and Representation: 103-111.

correct classifications (Figure 1). We have been studying this problem as part of our larger research goal of automatically extracting information from the web [Craven et al. 1998]. For example, one specific task we have considered is training a system to classify web pages into categories such as "student home page", "faculty home page", "home page of academic course", etc.

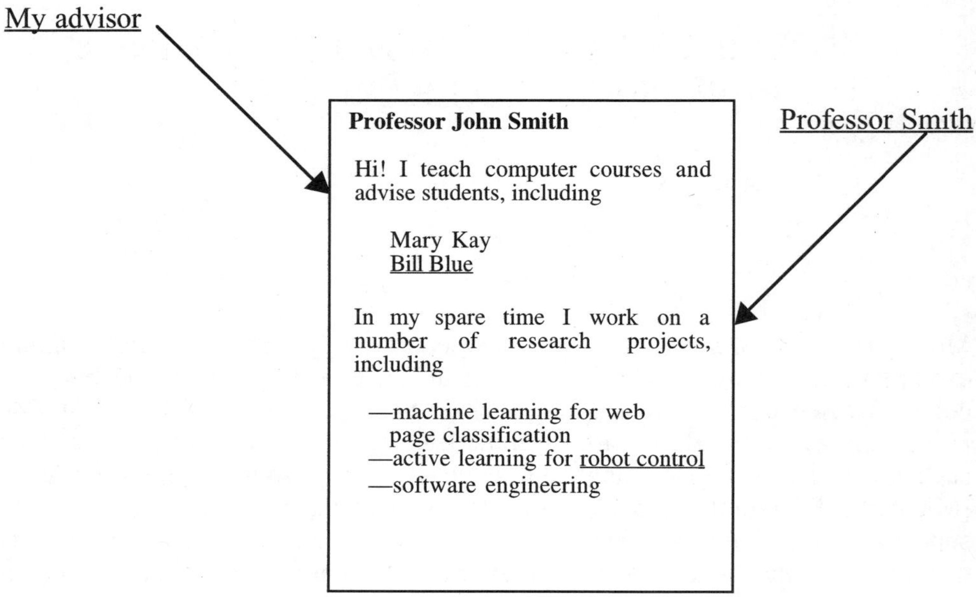

Figure 1: Training example of a "faculty home page." The task of classifying a web page can be achieved by considering just the words on the web page. Alternatively, the page can be classified using only the words on hyperlinks that point to the web page (e.g. "my advisor" "Professor Smith"). When examples are described by such redundantly sufficient features, *unlabeled data can be used to boost the accuracy for supervised learning.*

This web page classification problem can be formulated as a typical supervised learning problem. By "supervised learning problem" we mean that there is some set of X of instances, and there is some target function f: $X \rightarrow Y$ to be learned, given a set of training examples of the form $\{\langle x_i, f(x_i)\rangle\}$. In our case, X is the set of all web pages, Y is the set of possible classifications (e.g., "student" "faculty" "neither"), and the function f to be learned in the function that maps an arbitrary page x to its correct classification. For the ith labeled training example, x_i is a particular web page, and $f(x_i)$ is the correct classification for x_i, provided by an external teacher.

As discussed in [Craven et al. 1998], a number of supervised learning algorithms can be applied to the problem of learning to classify web pages. One common approach is to first represent each web page by a large feature vector, where each possible world in the language corresponds to a feature, and the value of the feature is the number of times the word occurs in the web page. This "bag of words" representation obviously ignores information about the sequence in which words occur. However, this representation of web pages as large feature vectors enables us to apply standard supervised learning algorithms for classification of features vectors. For example, if we train a naive Bayes classifier [Mitchell 1997] on a set of approximately 4,000 labeled web pages, we achieve accuracies of approximately 70% in classifying new web pages into the categories mentioned above [Craven et al., 1997].

Given:

- set L of labeled training examples
- set U of unlabeled examples

Loop:

- Learn hyperlink-based classifier H from L
- Learn full-text classifier F from L
- Allow H to label p positive and n negative examples from U
- Allow F to label p positive and n negative examples from U
- Add these self-labeled examples to L

Table 1: A Cotraining algorithm, for training two classifiers and using unlabeled data to boost their accuracies.

While we can achieve reasonable accuracy after training on thousands of hand labeled web pages, in this paper we are interested in the question of how unlabeled data can be useful. In fact, this is a crucial question in our web page classification task, because it is costly to hand label thousands of web pages, and because it is very easy to obtain hundreds of millions of unlabeled pages from the web.

How can we use unlabeled web pages to more accurately learn to classify future pages? The answer can be seen by first observing that the web contains an interesting kind of redundant information about each web page, as shown in figure 1. As shown in this figure, we could classify the web page either by considering the words on the page itself (as suggested in the paragraph above), or we could classify the page by ignoring the words on the page and instead considering the words on hyperlinks that point to the page. In many cases, the words on the hyperlinks will be sufficient to classify the example, and the words on the page itself will also be sufficient. In such cases, we will say these two sets of features — the hyperlink words, and the web page words — are *redundantly sufficient* to classify the example. We will see below that in general if the instances X for some supervised learning task can be factored into sets of redundantly sufficient features, then unlabeled data can be used to boots the accuracy of classifiers trained with limited labeled data.

3. LEARNING WITH REDUNDANTLY SUFFICIENT FEATURES

How can we use unlabeled data to boost the learning accuracy for our web classification problem?

	Page-based classifier	Hyperlink-based classifier	Combined classifier
Supervised training	12.9	12.4	11.1
Co-training	6.2	11.6	5.0

Table 2: Error rate in percent for classifyingweb pages as course home pages. The top row shows errors when training on only the labeled examples. Bottom row shows errors when co-training, using both labeled and unlabeled examples. The rightmost column shows the classification accuracy when the two classifiers are provided equal votes on the final classification. The results reported here are the average of five independent runs of the algorithm with different random starting examples.

The key idea is that we will train two independent classifiers rather than one. One classifier will use only the words on the web page, and the other classifier will use only the words on the hyperlinks. We begin by training both classifiers using whatever labeled training examples are available. Presumably this will result in two classifiers that are imperfect, but better than random. Now we use the unlabeled data as follows: each classifier is allowed to examine the unlabeled data and to pick its most confidently predicted positive and negative examples, and add these to the set of labeled examples. In other words, each classifier is allowed to augment the pool of labeled examples. Both classifiers are now retrained on this augmented set of labeled examples, and the process is repeated as long as desired. Table 1 summarizes this co-training algorithm.

Why should this co-training algorithm lead to more accurate classifiers? The intuition is that if the hyperlink classifier finds an "easily classified" hyperlink in the unlabeled data (e.g., one that is quite similar to one of the labeled examples on which it was trained), the web page that it points to will be added to the labeled pool of examples as well. Of course just because the hyperlink happened to be easy to classify does not mean the web page will be easily classified by the other classifier. If not, then the hyperlink classifier has added useful training information to improve the other classifier. Similarly, the web page classifier can add examples that are easy for it to classify, but that provide useful information to improve the accuracy of the hyperlink classifier.

In experiments, we have found that this Cotraining algorithm does improve classification accuracy when learning to classify web pages. In one experiment [Blum and Mitchell, 1998], summarized in Table 2, we trained a classifier to label web pages as home pages of academic courses. In this experiment we provided just 16 labeled examples, and approximately 800 unlabeled pages drawn from computer science department web sites. On each iteration of co-training, each classifier was allowed to add 1 new positive and 3 new negative examples to the pool of labeled examples. After 30 iterations of the cotraining algorithm, the accuracy of the combined classifier was 95%, compared to 89% when only the labeled data was used. In this case, the impact of cotraining was to reduce the error by better than a factor of two.

4. FORMAL RESULTS

Given the experimental evidence that cotraining can be useful in at least one case, it is useful to characterize the general problem setting in which this kind of algorithm can benefit from unlabeled data. Here we define a problem setting in which we can prove that unlabeled data can be of help.

To define the general problem setting that captures the essential structure of our web classification example, recall that earlier we defined a supervised learning problem as one where we have some set of instances X, a target function $f: X \rightarrow Y$, and labeled examples $\{\langle x_i, f(x_i)\rangle\}$. In the *cotraining* setting, we assume such a supervised learning problem in which the instances are drawn from X according to some fixed (possibly unknown) probability distribution D. We further assume that $X = X_1 \times X_2$; that is, the instance in X can be factored into two parts (e.g., the hyperlink words and the web page words). We require also that X_1 and X_2 each contain information sufficient to classify the example. In other words, we require that there exist some function g_1: $X_1 \rightarrow$Y and some function g_2: $X_2 \rightarrow$Y such that for all $x \in X$, $g_1(x_i) = g_2(x_2) = f(x)$. Note the learner is not expected to know f, g_1 or g_2, in advance. We simply require that it be possible to express f in terms of x_1 and in terms of x_2 (i.e., that X_1 and X_2 both be sufficient to classify X according to f). If the above constraints on X_1 and X_2 are satisfied, then we will say that X_1 and X_2 are *redundantly sufficient* to classify X with respect to f.

In [Blum and Mitchell 1998] we show that if f: $X \rightarrow Y$ is PAC learnable from noisy labeled data, X_1 and X_2 are redundantly sufficient to classify X with respect to f, and if x_{B1} and x_{B2} are conditionally independent given $f(x)$, then f can be PAC learned given a weak initial classifier plus only *unlabeled* data. What this means is that if the labeled data are sufficient to train a better than random initial classifier, if X_1 and X_2 are distributed independently for each target value of f, and if can be learned accurately given an arbitrary number of *labeled* examples, then f can be learned to similar accuracy given just this weak initial classifier and an arbitrary number of *unlabeled* examples. The significance is that this proves that unlabeled examples can substitute for labeled examples, under the conditions of the theorem.

In [Blum and Mitchell 1998] we also show that unlabeled data is of potential value under the more general setting in which X_1 and X_2 are not independently distributed. There we show that if the training data is noise-free, then unless X_1 and X_2 are deterministically related (e.g., one value of x_1 is always paired with the same x_2), the unlabeled data can be of use.

5. OTHER LEARNING TASKS WITH REDUNDANTLY SUFFICIENT FEATURES

Given the success of cotraining as a method of using unlabeled data when learning to classify web pages, and given the formal characterization of the class of problems for which cotraining can be of use, it is interesting to ask what other natural learning problems allow this approach. Here we summarize a number of tasks:

- Learning to classify noun phrases into semantic classes. In [Riloff and Jones 1999], an approach similar to cotraining was applied to the problem of learning a semantic lexicon over noun phrases in English. In this paper, one task was to learn to classify noun phrases as positive or negative examples of locations (e.g., "San Sebastian" is positive, "blue" is negative). In this case each example, x, is a sentence such as "We are located in lovely Pittsburgh". The factorization of x into two redundantly sufficient features sets is done as follows: x_1 is the noun phrase itself (e.g., "Pittsburgh" in the above sentence), and x_2 is the linguistic context in which the noun phrase appears (e.g., "We are located in lovely —"). Note that in most cases it is possible to determine whether the noun phrase is a location given either x_1 or x_2. Note also that the values x_1 and x_2 are distributed fairly independently (i.e., with might see the same x_2 as above, but with a different x_1). Starting with a list of just 10 known noun phrases, their system used an algorithm similar to cotraining to learn dozens of additional locations.

- Learning to select word sense. In [Yarowsky, 1995], an approach similar to cotraining was applied to learning to disambiguate word senses (e.g., to determine whether the word "plant" refers to a manufacturing plant or to a botanical plant). Here, each instance x corresponds to a context containing the word in question (similar to the Riloff and Jones example above). However, the factorization of x is into more than two components, with each word serving as a candidate component, and this algorithm seems to not map directly into cotraining. But it is quite similar, and the problem could be attacked using a cotraining approach.

- Learning to recognize phonemes in speech. In [de Sa and Ballard, 1998], an approach to fully unsupervised learning was applied to a problem with nearly redundantly sufficient features. In this case, the task is to learn to classify speech phonemes, based on both the audio signal and the video signal watching the speaker's lips. Here each instance x corresponds to the full data, x_1 corresponds to the audio signal, and x_2 corresponds to the video signal. Again, we may roughly assume that either of x_1 or x_2 is sufficient to determine the phoneme, and therefore can train two classifiers on limited labeled data, and let them train each other over the unlabeled data. In fact, de Sa and Ballard use no labeled data at all, instead performing a kind of clustering in which the audio signal must predict the video, and vice versa. Their system was able to learn clusters corresponding to the spoken phonemes in the data.

- Object recognition in multimedia data. Consider the problem of learning to recognize objects in multimedia data, such as the continuous stream of audio, video, and other sensory input a person receives. A very similar task is to learn to classify television segments, based on the ongoing stream of video, audio, and close captioned text. In this latter case, consider the problem of learning to spot television segments in which Boris Yeltsin appears. Here we could consider each instance x to be a snapshot containing the audio, video and text at a particular time. Here x_1 could be the audio, x_2 the video, and x_3 the text. Note that in some cases we might see "easy" example of video (a full face of Yeltsin), or an easy audio (his voice without background noise), or an easy text (the word Yeltsin). Thus, we could expect cotraining to provide a useful approach to using unlabeled data to train classifiers. To my knowledge, this experiment has not been attempted.

6. DISCUSSION AND CONCLUSIONS

We have described a class of supervised learning problems for which unlabeled data can be proven to improve learning accuracy. The key defining features of this problem class are that (1) the instances X can be factored into two or more components $X_1, X_2, \ldots$, which are *redundantly sufficient* to classify the example, and (2) these components covary, so that a particular x_i does not always co-occur with the same x_j.

We have described the cotraining algorithm that uses unlabeled data for such problems, and have presented experimental and theoretical results showing that in these problems unlabeled data can indeed be useful to improve accuracy for supervised learning.

The connection to human and animal learning appears to be a rich area for future study. Humans and other animals have a rich set of sensory input, which includes redundantly sufficient data for many tasks. For example, smell, vision, sound can all be useful in trying to classify whether food is nearby, and in many cases the data from just one of these channels is sufficient. de Sa and Ballard's success with a cotraining-like algorithm for learning to classify phonemes from video and audio — similar to that observed by humans — is also suggestive. Perhaps people rely less than we suspect on labeled data to achieve successful learning for various classification tasks.

ACKNOWLEDGMENTS

Thanks to my collaborator on this work, Avrim Blum, for many interesting discussions and ideas. Sebastian Thrun, Rosie Jones, Andrew McCallum, and Kamal Nigam have also contributed ideas and suggestions along the way. This work has been supported by Darpa under research contract F30602-97-1-0215.

Tom M. Mitchell
Carnegie Mellon University
U.S.A.

REFERENCES

Anderson et al. [1995], Production system models of complex cognition. In *Proceedings of the Seventeenth Annual Conference of the Cognitive Science Society* (pp. 9-12). Hillsdale, NJ: Lawrence Erlbaum Associates.

Blum and Mitchell [1998], Combining Labeled and Unlabeled Data with Co-Training, COLT98. Available at http://www.cs.cmu.edu/~webkb.

Craven et al. [1998], Learning to extract symbolic knowledge from the world wide web. In *Proceedings of the 15th National Conference on Artificial Intelligence (AAAI-98).* Available at http://www.cs.cmu.edu.

de Sa [1994], Learning classification with unlabeled data, NIPS-6, 1994.

de Sa and Ballard [1998], Category learning through multi-modality sensing, *Neural Computation* 10(5), 1998.

Riloff and Jones [1999], Learning dictionaries for information extraction by multi-level bootstrapping, *AAAI99.* Available at http://www.cs.cmu.edu/~webkb.

Mitchell [1997], *Machine learning*. New York: McGraw Hill, 1997. See http//www.cs.cmu.edu/~tom/mlbook.

Newell [1990], *Unified theories of cognition*. Cambridge, MA: Harvard University Press, 1990.

Yarowsky [1995], Unsupervised word sense disambiguation rivaling supervised methods. *Proceedings of the 33rd Annual Meeting of the ACL*, pp. 189-196.

JEAN-MICHEL ROY

COGNITIVE NEUROSCIENCE AND THE UNITY OF THE STUDY OF COGNITION

How the psychological and the neurobiological investigations of cognitive phenomena should be related within a materialist approach to cognitive science is a much debated issue, and one of central importance for the more general question of the unity of the scientific study of cognition. My ambition is here to engage in an inquiry into the significance for this issue of the recent developments of cognitive neuroscience, broadly construed as that part of contemporary cognitive research investigating the neurobiological dimension of natural cognition. Such an inquiry is a standard case of epistemological analysis, in the sense of an attempt to clarify what scientific knowledge should be on the basis of an examination of what it is in a given body of scientific work.

My concern will be mainly to lay the grounds for such an analysis. Accordingly, I will first (§1) attempt to provide a rigorous specification of what it means and of the main tasks it involves, as well as to account for its necessity. I will then (§2) introduce a general framework for carrying out a systematic investigation of the different relations that should hold between the neurobiological and the psychological levels of inquiry into cognitive phenomena from a materialist viewpoint. On the basis of these preliminary developments, I will finally (§3) sketch out some elements of an epistemological analysis of contemporary developments of cognitive neuroscience, with a view to clarifying what they teach the materialist about the appropriate way of relating these two levels.

Jesus M. Larrazabal and Luis A. Pérez Miranda (eds.), 113, Knowledge, and Representation: 113-133.

1. THE EPISTEMOLOGY OF COGNITIVE SCIENCE

1.1 The notion of epistemological analysis:

The epistemology of the sciences of cognition, which is now more than an emerging field, shares a number of features with the epistemology of other scientific disciplines. As a general matter, epistemology faces two main tasks.

The central one is commonly, and in my opinion correctly, described as that of determining whether and how scientific knowledge can be achieved, a goal which can be alternatively characterized as that of laying out the foundations of science. The foundational problems to be addressed by epistemology can be sorted out into two main categories, although the dividing line between them may at times look uncertain. On the one hand are formal problems, dealing with the nature of the operations and methods leading to the acquisition of a scientific kind of knowledge; on the other hand are what might by contrast be labeled material problems, dealing with the nature of the most fundamental concepts and principles to be applied to the elements under investigation. Material and formal foundational hypotheses can be formulated at different levels of generality.

Accordingly, the central task of the epistemology of cognitive science is to answer the following question: What are the adequate foundations of a science of cognition, that is, what are the principles on which it should be based *de jure* ? Whereas interest in the formal problems raised by such a question has remained peripheral, interest in the material ones has been, and remains, central for philosophers as well as for other members of the cognitive community. The issue of the appropriate way of relating the psychological and the neurobiological investigations of cognition figures prominently among these material problems.

The second main task of epistemology, which is a logical complement of the first, is to explore the implications, of domain-specific foundational hypotheses, as well as of the results obtained on their basis, for the understanding of the general nature of science. Here, the central question is: What is scientific knowledge if the investigation of such and such domain is founded upon such and such hypothesis and leads to such and such finding? It should be emphasized that the exploration of the general epistemological implications of a discipline can in principle induce readjustments in its own foundations.

In the specific case of cognition, this second task is concerned with laying bare the general epistemological consequences of adopting a certain perspective on the foundations of cognitive science, and of the results obtained in that perspective. The general epistemological implications of the results of cognitive science are of special importance, given the nature of its domain: it is indeed arguable that any general theory of science is based on a theory of cognition.

These two divisions of the epistemology of cognitive science (namely: the investigation of the *de jure* foundations of cognitive theories and the analysis of the implications of these foundations for the general conception of scientific knowledge) correspond to what Ronald Giere has proposed to label respectively the "philosophy of cognitive science" and "the cognitive philosophy of science"[1].

Epistemological analysis can in turn be defined as a further element of epistemology that is specifically concerned with the interpretation and assessment of already developed scientific theories, with respect both to their foundations and to their general epistemological implications. Accordingly, it cuts across the divisions of epistemology just mentioned. On the one hand, it tries to answer the following questions: a) what are *de facto* the foundations of cognitive science? and b) are these *de facto* foundations legitimate ones, that is, are these the foundations that cognitive science should receive *de jure*? On the other hand, it clarifies and evaluates the general epistemological implications of the foundations *de facto* assumed by cognitive science, as well of the results obtained on their basis.

From a foundational point of view, the need for an epistemological analysis derives mainly from two basic considerations. The first is that the choice of a *de jure* foundational hypothesis cannot be rational unless that hypothesis is shown to be different from and better than competing ones. Hence the need for a clarification and an assessment of the *de facto* foundations of empirically developed cognitive theories, a critical inquiry also of heuristic value for formulating an hypothesis about *de jure* foundations. The second consideration is that the justification of a foundational hypothesis is largely dependent upon the empirical adequacy of its logical consequences. Hence the need for a clarification and an assessment of the foundational implications of the empirical results obtained by already developed cognitive theories, and, more specifically, of the consistency of these results with the foundational hypotheses that are supposed to support them [2]. In the case of cognitive science, the clearest illustration of such necessity is provided by the issue of reductionism: If the reduction of psychology to neurobiology presupposes one-to-one nomic correlations between psychological properties and neurobiological ones, then a reductionist hypothesis cannot be acceptable unless available empirical research makes such correlations probable.

The overall resulting architecture of the epistemology of cognitive science can be schematized as follows:

Figure 1

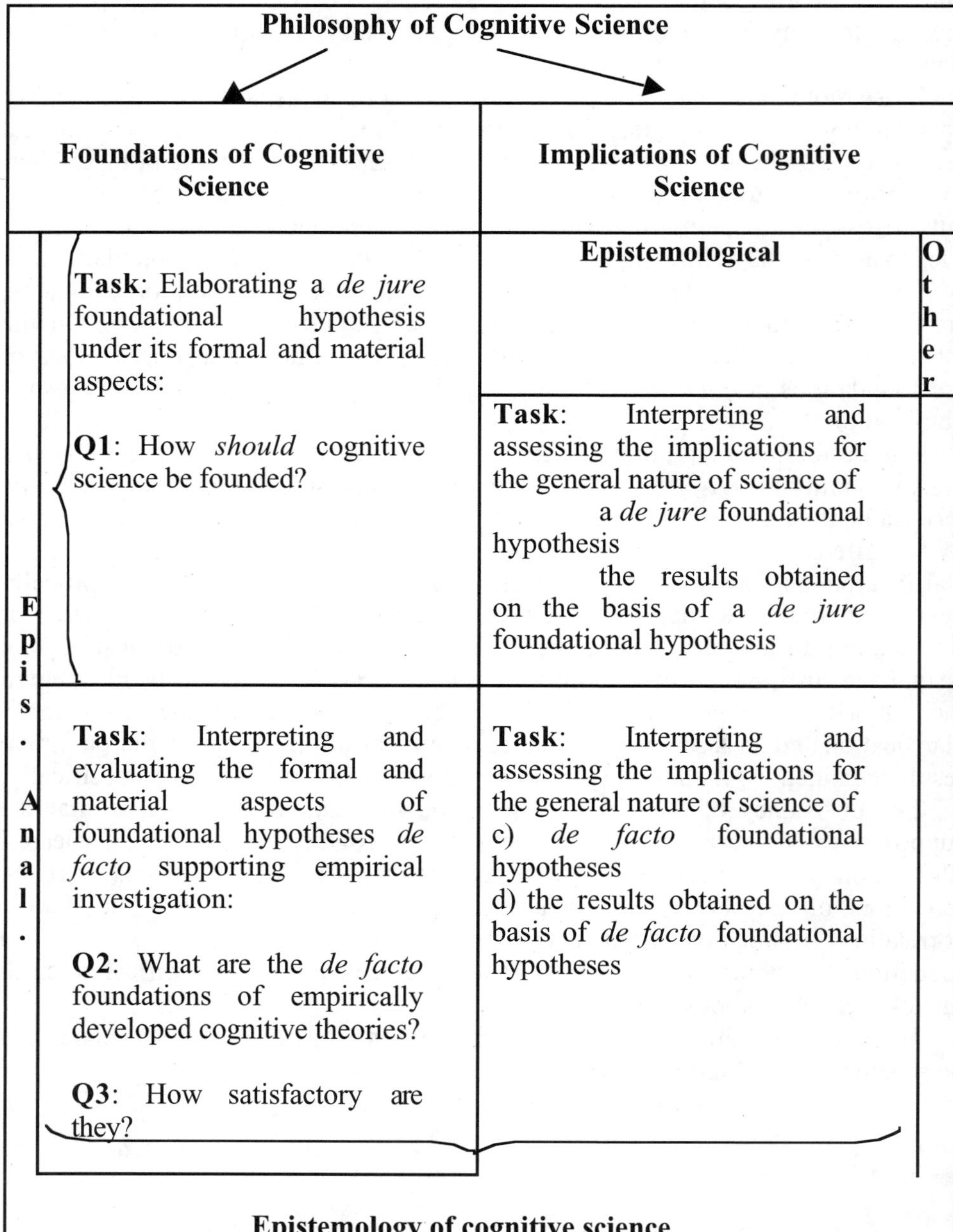

As a consequence, the study of any foundational problem of cognitive science should typically include at least three main steps.

First, it must determine what solution - if any [3] - empirically developed theories provide to the problem under consideration (Q2).

Second, it must then assess the adequacy of that solution (Q3), including by testing its consistency with the empirical results correlated with it. These first two steps belong to epistemological analysis.

Finally, a new foundational hypothesis has to be formulated (Q1). This third step is necessarily distinct from the second one. Indeed, even in the case where the solution *de facto* adopted by empirically developed theories is deemed adequate, it will inevitably call for a reformulation, and even at times for a clarification, as well as for the introduction of a number of modifications.

A fourth additional step, also of a critical nature, is in fact crucially needed. Indeed, the new hypothesis obtained about the foundations of cognitive science must be confronted with competing *de jure* hypotheses, a confrontation including, again, a clarification of their differences and similarities in theoretical content and an evaluation of their respective merits. Now, assuming that each one of these competing hypotheses is itself grounded on an epistemological analysis, the confrontation should also bear upon the differences in the interpretation and evaluation of empirically developed theories put forward by these analyses. In other words, in such a case the confrontation should be conducted at the level not only of Q1, but also of Q2 and Q3. And this task also belongs to epistemological analysis as previously defined.

1.2 The need for an epistemological analysis of cognitive neuroscience:

The problem of the relation between the psychological and the neurobiological investigations of cognition is of a foundational nature, and therefore no exception to the general schema of investigation just indicated. Accordingly, its solution requires an epistemological analysis of empirical research in cognitive science, and in cognitive neuroscience in particular.

There are two main reasons for paying special attention to cognitive neuroscience research. One is its dynamism, a dynamism unanimously emphasized by all sides of the cognitive community and sustained by a number of technical breakthroughs magnifying its power of observation, and as a result of which it was pushed from the background to the foreground of cognitive science in a matter of less than thirty years. Indeed, the development of a scientific discipline always carries with it some clarifications and some modifications of its foundational aspects, both of which have the potential to transform the difficulties raised at an earlier stage of its evolution. The second reason is that crucial aspects of the materialist solutions advocated about the relations between the psychological and the neurobiological investigations of cognition explicitly claim empirical grounds, and are thereby made *ipso facto* dependent upon the evolution of neurocognitive research. In a period of exceptional growth of this research, - that is, when conjugated with the previous one - this second reason becomes an even more pressing incentive to examine whether the wind of current neurocognitive investigation is blowing into the direction of one of the predicted solutions or is taking some unforeseen path.

In view of the developments of the previous paragraph (§1.1), the main goal of such an epistemological analysis is to put into light what conception of these relations serves *de facto* as a foundation for neurocognitive research (cf. step 1/Q2) and to assess its value (cf. step 2/Q3), particularly by investigating its consistency with the empirical results obtained.

It must also anticipate the confrontation of the *de jure* hypothesis which can be elaborated on the basis of these first these two tasks, with competing *de jure* hypotheses that are themselves derived from an epistemological analysis of empirical cognitive research, and of neurocognitive research in particular (cf. step 4). As a matter of fact, a number of contemporary dissenting theories clearly qualify as such hypotheses, in the sense that each of them offers a more or less complete materialist view of how cognitive science *should* be founded – and, more specifically, should relate the psychological and the neurobiological levels of investigation - and claims at the same time, although to a different extent, to be based upon an epistemological analysis of empirical cognitive neuroscience research.

It is convenient to approach this materialist foundational debate with the standard categories of eliminativism, reductionism, and functionalist as well as emergentist non-reductionism, even though such a classification only refers to the ontological aspect of the relation between the psychological and neurobiological investigations of cognition. Strict eliminativism is commonly attributed to the views expressed by Jean-Pierre Changeux in *L'homme Neuronal* [4], in spite of the fact that they might better be read as incarnating a typically reductionist approach. Neurobiological reductionism finds a clear expression under the name of computational neuroscience or neurophilosophy in the work of Terence Sejnowski, and Patricia and Paul Churchland.[5] Finally, non-reductionist materialism corresponds to the stand taken by classical information processing psychology and the philosophical exegesis of it provided by Jerry Fodor or Zenon Pylyshyn, although functional versions of a non-computational kind, as well as emergentist versions, have also been defended.

It is my contention, however, that there is another representative of non-reductionism which is of special importance for the confrontation under consideration. Indeed, this non-reductionist hypothesis is entirely the product of practicing cognitive neuroscientists and is a direct outcome of an epistemological analysis of the transformations recently accomplished in their discipline. For these reasons, it is unfamiliar to philosophers and is simply referred to as cognitive neuroscience. This expression is therefore ambiguous: It is used both in a generic and in a specific sense. In its first sense, it merely designates the investigation of the role played by the nervous system in cognition and is synonymous with the notion of neurobiology of cognition. In *Cognitive*

Neuroscience, Michael Gazzaniga traces the origins of that first use of the expression back to the late 1970's [6]. In its second sense, it refers to a specific hypothesis about the foundations of cognitive science, mainly characterized by the view that neurobiological research should play a fundamental role in cognitive theory. Michael Gazzaniga himself can be seen as one of its leading representatives. Indeed, as well illustrated in the opening chapter of the same book, he not only promotes the development of the neurobiology of cognition, but also defends a specific epistemological analysis of its recent achievements, claiming that they amount to a substantial and more adequate reorganization of the internal architecture of cognitive science, and in particular of the relation between the psychological and neurobiological investigations of cognition. Steve Kosslyn and Olivier Koenig are even better representatives of this trend with their 1992 book, *Wet mind: The New Cognitive Neuroscience* [7], where they try to sharply dissociate the way cognitive neuroscientists relate the study of the brain to the study of the mind from the way advocated by classical information-processing psychologists. Michel Imbert [8], Irvin Rock [9], Peter Eimas [10], Alain Berthoz [11], Michael Posner [12] and several other authors share similar views. It should not be concluded that this epistemological analysis and the *de jure* hypothesis derived from it have yet found a satisfactory articulation in terms of precision and homogeneity. Nevertheless, they offer sufficiently characteristic features and make sufficiently consistent claims of originality to be counted as a distinctive element of the materialist debate, and to be labelled with a capitalized Cognitive Neuroscience.[13]

In addition to offering an interpretation and assessment of the foundations *de facto* assumed by empirical neurocognitive research, the epistemological analysis advocated here must therefore also a) determine the extent to which these main representatives of the materialist foundational debate are effectively based themselves upon similar analyses; b) provide a detailed characterization of the contents of these analyses; and c) assess their own adequacy.1.3 Limitations:
The program of investigation just delineated clearly extends beyond the material limits of this essay, and my developments will have to remain at a preliminary level; setting up a detailed and rigorous research agenda is however an important task. I will therefore restrict my ambition to outlining an interpretation and assessment of the position *de facto* assumed by empirical neurocognitive research on the issue of inter-level relations, leaving entirely aside the confrontational aspect of the program [14]. In terms of the division of tasks previously introduced, I pretend to offer no more than a first try at steps 1 and 2 of the required epistemological analysis. Paradoxically and for similar reasons of space, I will also deliberately present the main results of this analysis in an abstract manner, summarizing them as a series of eleven hypotheses and abstaining from going into the details of the neuroscientific literature.

2. THE LOGICAL SPACE OF THE MATERIALIST APPROACH

This restricted inquiry presupposes a general conception of what counts as a materialist analysis of the relation between the psychological and the neurobiological studies of cognition that should however be first clarified.

By definition, its most basic assumption is clearly that cognitive science sees it as relevant to make use of two classes of apparently heterogeneous properties, the psychological and the neurobiological, along with two *prima facie* different types of investigations, formulating what might be called respectively H_P and H_N hypotheses about these properties. Second, it also assumes that, in order for the relation *R* between H_P and H_N to be a materialist one, it is necessary that H_P hypotheses attribute the psychological properties to the same entities that are attributed the neurobiological properties in H_N hypotheses. In addition, the attributions contained in H_P hypotheses should be capable of a rational explanation (and thereby of a rational justification) that only makes use of physical properties. Physical properties are understood as the class of properties accepted by the physical sciences at a time *t* of their development, as well as those which can be explained on their basis, and which include the neurobiological properties. In other words, an H_P is related to an H_N in a materialistic way if it attributes a psychological property to the same x to which the H_N attributes a neurobiological property, and if that attribution can be rationally explained (and thereby justified) either by the neurobiological properties explicitly mentioned in H_N, or by physical properties presupposed by H_N.

A first crucial factor of specification of this general definition lies in the relations of dependence that can be introduced between H_P and H_N hypotheses. As a general matter, there are several possible relations of dependence between two hypotheses belonging to different disciplines. While some of them are necessary correlates of the materialist definition of *R*, others simply represent optional but fundamental sources of variation of such definition.

Ontological dependence between the properties studied by each one of the two psychological and neurobiological hypotheses is a first important type. This ontological dependence is variable in strength. In its minimal form, it only bears upon the existence of the properties, in the sense that a psychological property can only be instantiated in an entity if that entity instantiates a physical property, although any psychological property can be co-instantiated with any physical property. In such a case, there is no possible materialist account of the nature of the psychological properties. At the other end of the spectrum, the ontological dependence can also be such that the instantiation of a definite kind of psychological property requires the instantiation of a definite kind of neurobiological property. A full materialist account of psychological properties is then possible. The scope of the materialist explanation is thus proportional to the strength of the ontological dependence. Accordingly, a materialist definition of *R* is clearly inseparable from introducing such a dependence of psychological properties on neurobiological, and ultimately physical, properties. The former can only be explained by the latter if they are ontologically dependent on them.

Ontological dependence between the existence or the nature of properties is to be distinguished from different kinds of relations of dependence that can obtain between the corresponding attributing hypotheses, and can, for that reason, be grouped under the general heading of epistemological dependence. In a general manner, a body of scientific knowledge is dependent upon another one when the obtaining of the latter is a condition of the obtaining of the former. Such dependence can affect all main aspects of scientific knowledge, like its epistemological form, its methodology, the definition of its central problems, its evidential basis, and most of all, its theoretical content, to be broadly understood as the set of answers given to the problems investigated. Theoretical dependence is therefore the central form of epistemological dependence. It is important to further distinguish between three sorts of theoretical dependence, which I will respectively call logical, epistemic and heuristic.

A scientific hypothesis is logically dependent upon another one when its truth is somehow made dependent on the truth of that other hypothesis. Heuristic and epistemic dependencies, on the other hand, have to do with the elaboration of scientific knowledge. The first one belongs to the order of discovery: A scientific hypothesis is heuristically dependent upon another one when its formulation is dependent upon the previous formulation of that other hypothesis. The second one belongs to the order of justification: A scientific hypothesis is epistemically dependent upon another one when the determination of its truth-value cannot be obtained prior to that of the other hypothesis.

The logical dependence of H_P on H_N is a necessary correlate of the explanatory requirement contained in the materialist definition of *R*, as well as of the ontological dependence presupposed by such definition. If the attribution of a psychological property is explained by the attribution of a neurobiological one, the first one cannot be true unless the second one is also true. Similarly, if a psychological property cannot be instantiated unless a neurobiological property is instantiated, the hypothesis that the first one is instantiated cannot be true unless the hypothesis that the second one is instantiated is also true. The logical dependence of H_P on H_N is therefore proportional to the ontological dependence of psychological properties on neurobiological properties. However, an hypothesis can be logically dependent on another even though there is no explanatory relation between them, nor any ontological dependence between the properties they are dealing with [15]. As a consequence, the materialist definition of R does not require, but is compatible with, the idea that H_N is logically dependent on H_P.

Finally, the materialist definition of *R* is not committed to any specific form of heuristic or epistemic dependence. Indeed, the fact that an H_P should be explained by an H_N does not require that it cannot be discovered independently and justified on fully independent grounds. It is even conceivable that H_N be heuristically and epistemically dependent on H_P in a materialist perspective on *R*. Heuristic and epistemic dependencies vary independently of ontological and logical dependencies.

A second crucial factor of specification of the materialist conception of *R* is the inclusion or the rejection of an identity relation between the psychological properties mentioned in H_P and the neurobiological properties mentioned in H_N hypotheses. This is the alternative lying behind the standard opposition between reductionism and non-reductionism.

Indeed, as clearly illustrated by E. Nagel's standard definition [16], classical psycho-neural reductionism may be defined as the view that psychological hypotheses can be derived deductively from neurobiological ones (ultimately, from purely physical ones), and consequently merged with them into a single theory. When supplemented with the nomologico-deductive conception of scientific explanation [17], this reductionist picture is equivalent to saying that psychological hypotheses can be explained in terms of neurobiological ones by being deductively derived from them. However, such derivation requires that the psychological hypotheses be reformulated in neurobiological terms, given the heterogeneity of the properties mentioned in each of them, a requirement labeled by Nagel the "condition of connectability". This condition is fulfilled by establishing rules of correspondence between psychological properties and new neurobiological properties, obtained by way of logical construction from previously established ones. There are different ways of interpreting these rules of correspondence and this is a crucial issue for the definition of reductionism. The most standard one consists in considering them as identifications of two properties occurring in *prima facie* distinct nomic hypotheses. This line of interpretation opens up a number of additional alternatives, and in particular the choice between an extensional and an intensional construal of identity statements, as well as between an analytic (and *a priori*) or synthetic (and *a posteriori*) kind of justification for them. The typical view incarnated by the late 1950's identity theory that served as the main target of functionalist non-reductionism is both extensional and synthetic. More specifically, it is based on the hypothesis that, for any instantiation of every psychological property that can be mentioned in a nomic hypothesis, there is a correlated instantiation of a specific neurobiological logical construct; and therefore, that the two properties can be considered as true of the same things or extensionally identical, thereby opening the possibility for the psychological hypothesis to be reformulated as a neurobiological one, and eventually integrated in the deductive and explanatory framework of neurobiological theory (and ultimately of physical theory).

By specifying the general requirements that a general theory of *R* should meet in order to qualify as a materialist one, as well as the main options it faces in terms of relations of theoretical dependence and ontological identity, the previous developments somehow delineate the logical space of the problem of such a materialist theory. A logical space that can be usefully represented with the following schema, where the black signs stand for necessary conditions and the gray signs for optional ones:

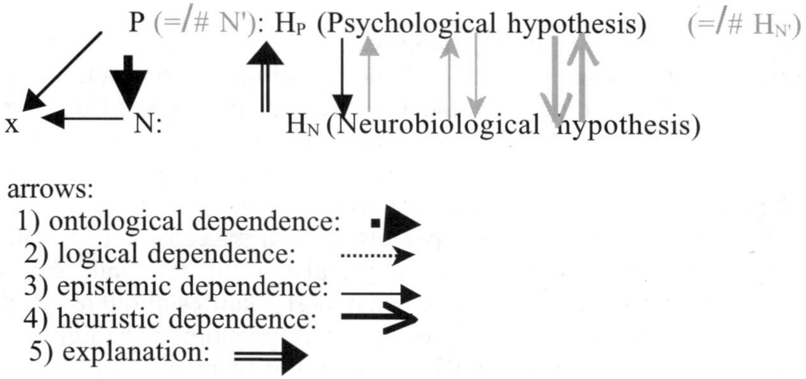

Figure 2

3. A SKETCH OF AN EPISTEMOLOGICAL ANALYSIS

In sketching out an epistemological analysis of contemporary cognitive neuroscience, it is important to avoid the temptation of oversimplification and to respect the diversity of epistemological features exhibited in the course of actual investigation. Nevertheless a number of characteristics clearly emerge with respect to the issue under consideration.

1) Criteria of inclusion into cognitive neuroscience:

Firstly, inclusion into the category of cognitive neuroscience seems to obey a couple of general criteria. On the one hand, an explanation of nervous activity only counts as neuroscientific if it explicitly mentions anatomical and physiological properties, thereby ruling out purely functional explanations, which, in a materialist perspective, are to be understood as explanations of the nervous system. On the other hand, a neuroscientific explanation of the nervous system is to be counted as a cognitive one only if it somehow relates nervous activity to something considered as cognitive activity. Thus, a study of the mechanisms of synaptic transmission or of action potential propagation does not by itself count as an element of cognitive neuroscience research. It does, however, if it is conducted in connection with the elucidation of a cognitive process or property, or the moment when it is reinvested in the context of such an elucidation. By contrast, the minute analysis of the connectivity of retinian cells would be spontaneously included into cognitive neuroscience, for the simple reason that it is naturally seen as an element of a comprehensive analysis of what the nervous system does when accomplishing the cognitive activity of visual perception. The typical form of the question being investigated by a cognitive neuroscience inquiry can consequently be put as follows: what does the nervous system do when such and such cognitive phenomenon is taking place?

2) A commitment to two types of properties:
The characterization of the cognitive phenomenon under consideration can be of three main sorts. First, it may exclude any reference to a type of property that is both internal and abstract. By abstract, I mean here non-neurobiological and non-physical. This form of characterization is frequent in the area of motor control, where the focus of inquiry is either on purely behavioral manifestations such as arm-reaching, hand-grasping, gait, posture, ocular saccadic movements, or on purely behavioral manifestations and physical stimuli. Second, the cognitive phenomenon may be described as a behavioral manifestation associated with abstract internal properties. Finally, the cognitive phenomenon might be reduced to such properties. The Pet or FMRI investigation of mental imagery tasks like simulated navigation offers a perfect illustration of that third case. The internal abstract properties are usually considered as psychological ones, as well as designated with familiar terms like those of intention, memory, judgment… For convenience, I will hereafter refer to them as psychological properties. However, it is important to emphasize that nothing crucial hinges on their being considered psychological, or also on their being assimilated with the referents of ordinary mental terms.

The use of psychological properties (in the sense of abstract and internal properties) in the characterization of the cognitive phenomena being investigated certainly is the hallmark of contemporary cognitive neuroscience, thereby showing that it is deeply committed to the admission of an additional type of property beyond the neurobiological one. Accordingly, its central interrogation can be more specifically rephrased as follows: What does the nervous system do when such and such psychological property is being instantiated ? This question is also often cast in terms of neural basis: What are the neural bases of the instantiation of such and such psychological property? However, the reference to psychological properties is not always made explicit in the formulation of the specific problem being investigated. Psychological properties may, for instance, simply be introduced in the course of what is initially presented as a search for the explanation of a purely neurobiological phenomenon. At times, this introduction may even fail to be recognized as a way of resorting to a non neurobiological kind of property.

3) Status of psychological properties:
A question naturally arises as to the status of the psychological hypotheses which are consequently involved in cognitive neuroscience research: Do they belong to cognitive neuroscience itself or are they just borrowed from another

field? In other words: Is cognitive neuroscience simply trying to figure out the neural bases of the psychological properties attributed to the nervous system by whatever discipline has the function of attributing them, or does it itself pretend to fulfill that function? This is a problem about the division of labor between the discovery of psychological properties and that of their neural bases. Some cognitive neuroscientists, like Kosslyn and Gazzaniga, claim that both tasks are the lot of cognitive neuroscience. For the most part, however, empirical research remains silent on the issue. Nevertheless, one might reasonably say that it also acts for the most part as if the two investigations were thought to belong to two different disciplines.

Where, then, are psychological hypotheses borrowed from? Contemporary neurocognitive research apparently makes use of three main sources: scientific cognitive psychology, a variety of intellectual theories such as philosophical ones, and finally sheer common sense or folk psychology. The first case is well illustrated by so-called "combined experiments" [18]. The last one of these three possibilities is of course highly problematic, though it might be the dominant one. Indeed, a cognitive neuroscientific explanation, being concerned with exhibiting the neural bases of a cognitive phenomenon, can fail either by providing the wrong bases, or by searching for the bases of an inexistent, misinterpreted, or ill-defined cognitive phenomenon. By resorting to folk psychology, one is clearly incurring the risk of being the victim of the last one of these two kinds of error. In any case, reliance on commonsense seriously affects the scientificity of the explanation proposed since, if psychological commonsense may be true (as Fodor strongly believes), it must still be shown to be true. Moreover, even in the eyes of its most zealous advocates, the principles of commonsense also suffer from numerous approximations.

4) Ontological dependence of psychological properties on neurobiological properties:

The notion of basis, as well as that of substrate also in common use, both indicate something upon which the existence and nature of psychological properties are dependent. Indeed, these notions imply that the instantiation of psychological properties is supported by the instantiation of neurobiological properties. Accordingly, cognitive neuroscience unambiguously introduces a relation of ontological dependence of psychological properties on neurobiological properties.

It is correct, however, that the more neutral term of neural correlate is sometimes substituted for those of neural basis and neural substrate. The debate raised by F. Crick [19] around the role of V1 in consciousness is a good case in point. *Stricto sensu* the notion of correlate only implies that a psychological property cannot be instantiated without the corresponding instantiation of a neurobiological property, but does not establish any ontological priority between the two instantiations, whereas the notion of basis involves the idea of a necessary but asymetrical co-instantiation. However, it is far from clear that, except in a few cases, the recourse to a neural correlate terminology is intended to bring out such a conceptual difference and that it should therefore be taken at face value.

The difficulty lies less in the admission by cognitive neuroscience research of such an ontological dependence, than in the force attributed to it. The issue is actually explicitly debated within cognitive neuroscience in relation with the problem of neural plasticity, that is, of the actual degree of variation in the neural correlates of psychological properties. Depending on the stand adopted on this problem, the ontological dependence is made more or less strong. It should at any rate be emphasized that neural plasticity is standardly considered to be a restricted phenomenon, if a more pervasive one that it was once thought [20]: cognitive neuroscience has it that evolution did not make human brains wildly different from individual to individual, for the alternative would be antagonistic to the very project of a cognitive neuroscience.

5) Materialist commitment and ontological neutrality:

Is cognitive neuroscience research subscribing to materialism by accepting that psychological properties are ontologically dependent upon neurobiological properties? A materialist conception of *R* has been defined above as a conjunction of two theses (1° attributing P properties to the same entity that is attributed N properties; 2° explaining P properties by means of N properties), and it has also been proposed that the second one of them entails the admission of such ontological dependence. The reverse implication also holds: if it is admitted that P properties are ontologically dependent upon N properties, then it is admitted that the instantiation of P properties is to be explained by means of the instantiation of N properties. However, the first thesis contained in the materialist definition of R is not thereby necessarily satisfied. Indeed, it is in principle possible that the instantiation of psychological properties depends upon the instantiation of neurobiological ones (ontological dependence), even if both kinds of properties are not instantiated in the same entity. However, by making the instantiation of neurobiological properties a condition for the instantiation of psychological properties, cognitive neuroscience looks clearly animated by the materialist intention, not only of providing explanations of the psychological attributions in terms of neurobiological ones, but also of attributing the former properties to the very entity that is endowed with the latter ones. What cognitive neuroscience is undoubtedly after is an explanation of *how* the nervous system *itself* succeeds in being a psychological system. Here again, even when the less ontologically committing term of correlation is used, it is far from clear that the substitution reflects a difference in ontological perspective.

However, that term probably gives a more adequate expression of what the ontological attitude of cognitive neuroscience research really comes down to. For its real achievements are mainly limited to establishing that the instantiation of specific psychological properties is inseparable from the instantiation of more or less specific neurobiological properties. In other words, in most cases, it does not really go beyond establishing nomic correlations between such instantiations.

Indeed, the materialist claims of cognitive neuroscientists are frequently not up to their ambitions on a number of points. Only on rare occasions do cognitive neuroscientists attribute in an accurate and explicit way the psychological properties to the very subject of the neurobiological ones. And when they do, they usually do not offer any justification, let alone any explanation, for these attributions. Finally, even when they do attempt to provide such explanations, their efforts usually fall short in an important respect. The explanatory link advocated between the psychological analysis and the neurobiological analysis of the cognitive process remains unsatisfactory. And this explanatory gap is of course not only a matter of details, but also and more crucially one of principle. What is usually missing is a clear statement of how the type of property invoked in psychological explanations can be accounted for by the type of property invoked in neurobiological explanations. What is usually missing, in other words, is a principle of naturalization, thereby leaving the explanatory gap unfilled at a very fundamental level.

Accordingly, cognitive neuroscience research should either refer explicitly to some of the general solutions to the naturalization problem advocated by philosophers, whose main business in this area over almost a century has been to discover a principle for bridging that gap, or propose one of its own. But cognitive neuroscience research, with very few exceptions, is conspicuously silent on the issue. The psychological properties it needs to postulate are conceived as properties belonging to the same entity as their neural bases, but it is not not shown how they can do so. And even when a reference to one of the available solutions to the naturalization problem, such as reductionism or emergentism, is explicitly introduced, it remains in general dogmatic and non-technical.

A very large part of cognitive neuroscience research is therefore in a paradoxical situation. On the one hand, it is definitely animated by a materialist intention, to the point that its occasional use of a non-materialist vocabulary, such as the term "correlate", should not be taken at face value. On the other hand, its materialist position remains for the most part unjustified, and consequently without real substance, to the point that its occasional use of the term "correlate" should indeed be taken as expressing its real ontological significance. And that significance can be fairly summarized by saying that most results of cognitive neuroscience are ontologically neutral behind a materialist façade.

6) Logical dependence of psychological hypotheses upon neurobiological hypotheses:

As previously mentioned, the admission of a relation of ontological dependence between two properties requires the admission of a parallel relation of logical dependence between the corresponding hypotheses. Accordingly, by assuming the form of a search for the neural bases of psychological properties, cognitive neuroscience is committed to the following idea: the proposition that a specific psychological property is instantiated can be true only if the proposition that a more or less specific neurobiological property is also true.

7) Heuristic independence of psychological hypotheses from neurobiological hypotheses:

Is the relation of logical dependence just mentioned complemented with a heuristic one? As also indicated before, from the fact that an hypothesis H_1 is true only if an hypothesis H_2 is true, it does not follow that H_1 cannot be formulated independently of H_2. And as a matter of empirical research, the psychological hypotheses whose neural bases are being searched for in cognitive neuroscience investigations are for the most part purely behavior-based, and not brain-based, at least in any significant way. Here, it is important not to confuse the process of *confirming* an hypothesis about the neural bases of a psychological hypothesis, with the process of *elaborating* a psychological hypothesis that is heuristically brain-based.

This *de facto* situation raises the important question whether it is the result of a theoretical impossibility. In other words, is it impossible to derive the knowledge of psychological properties from the knowledge of their neurobiological conditions of instantiation, or, to use Gazzaniga's terminology, to understand which mind is enabled by the brain through inspection of the brain ? Standard practice in cognitive neuroscience suggests that it is, without explaining why. If it is not, that standard practice might just be one among other possible ways of searching for the neural bases of psychological properties of cognition, and therefore not the best, as the advocates of Cognitive Neuroscience and neurophilosophy actually claim.

8) *Epistemic dependence of psychological hypotheses on neurobiological hypotheses*

From the fact that an hypothesis H_1 is true only if an hypothesis H_2 is true, it also does not follow that one cannot have reasons to believe in the truth of H_1 that are fully independent from the reasons justifying the belief that H_2 is true.

And indeed the psychological hypotheses used by cognitive neuroscience are clearly justified, in large part, by non-neurobiological evidence, such as purely behavioral or introspective evidence [21]. Nevertheless, the situation in cognitive neuroscience research is also clearly one of limited epistemic dependence of H_P on H_N, since it is assumed that neurobiological hypotheses have an independent and specific evidential basis, and consequently, that psychological hypotheses cannot be fully established until neurobiological hypotheses themselves are. If an

H_P hypothesis is logically dependent on an H_N hypothesis, and if H_N is epistemologically independent, at least partially, I cannot know whether H_P is true until I know whether H_N is true.

9) Heuristic dependence of neurobiological hypotheses on psychological hypotheses:
However, what is probably the most striking, though barely discussed [22], feature of cognitive neuroscience research in terms of theoretical dependence is that it introduces a heuristic dependence of neurobiological investigation upon psychological investigation. Indeed, in searching for the neural bases *of* psychological properties, cognitive neuroscience conducts the study of neurobiological properties in the light of previously obtained knowledge of psychological properties: the neurobiological inquiry leans against the psychological hypothesis. And here again, it is important to determine to what extent such dependence is necessary, and why. Standard practice in cognitive neuroscience suggests that it is necessary; but if it is not, then such practice corresponds to no more than one possible way of conducting neurocognitive research, and it might again not be the most adequate.

10) Epistemic dependence of neurobiological hypotheses on psychological hypotheses:
To what extent is the heuristic dependence just emphasized complemented with an epistemic one? It was already acknowledged that neurobiological hypotheses are justified on independent grounds, since they are clearly based on specific data. It is nevertheless disputable that their adequacy is fully assessable independently of that of psychological hypotheses. They certainly are independent from psychological hypotheses to the extent that they are about neurobiological properties. But being hypotheses about the neural bases *of* psychological properties, they are better seen as implicitly relational statements: they assert that certain neurobiological properties are instantiated *when* certain psychological properties are instantiated. And as such, they cannot be evaluated independently of psychological hypotheses. We cannot know whether they are true until we know whether the corresponding psychological hypotheses are themselves true.

11) Logical dependence of neurobiological hypotheses on psychological hypotheses:
Finally, for the same reason, neurobiological hypotheses should be seen as being partially dependent on psychological hypotheses from a logical point of view. Just as a neurobiological hypothesis cannot be known to be true, unless the psychological hypothesis with which it is intrinsically related in virtue of the very form assumed by the search for neural bases, is known to be true, so too

such a neurobiological hypothesis cannot be true unless that psychological hypothesis is true.

4. CONCLUSION

Contemporary cognitive neuroscience research seems *de facto* founded upon a largely prevailing form of materialist analysis of the relation *R*, which can usefully be interpreted as the result of fixing some of the variables mentioned in figure 2. Consequently, it can be itself schematized as follows:

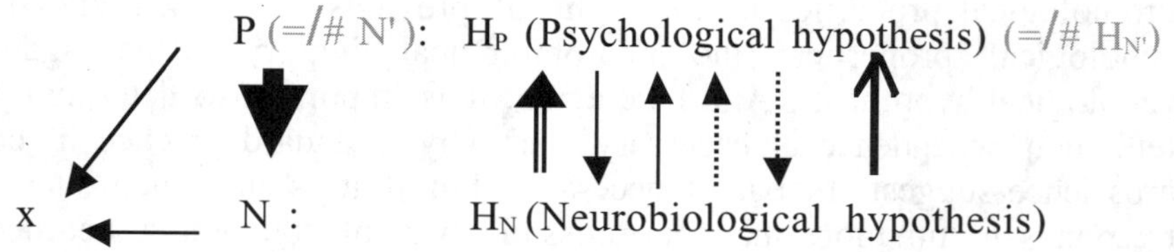

Figure 3

The remaining gray elements are meant to reflect the lack of a clearly dominant position on the issue of reductionism, and, more generally, the lack of a clear stand on the nature of the principle of naturalization.

If, in spite of an unavoidable element of simplification, it is indeed adequate to the facts, the epistemological analysis just sketched out reveals that contemporary cognitive neuroscience research is supported by a rather well-delineated foundational hypothesis, although one suffering from a number of deficiencies. Three of these deficiencies, of special importance, have been pointed out.

In the first place, some of the established relations of theoretical dependence and independence stand in crucial need of justification, in view either of their own apparent implausibility or of the apparent plausibility of alternative choices. Such are the respective cases of the heuristic dependence of neurobiological hypotheses on psychological hypotheses and its opposite, the heuristic independence of psychological hypotheses from neurobiological hypotheses. The reciprocity of the logical and epistemic dependencies associated with them also calls for a clarification because of their apparently circular character.

In the second place, it was emphasized that the scientific status of the psychological hypotheses serving as a heuristic, epistemic and logical foundation of the investigation of neural bases is often controversial. Reliance on commonsense as such is not acceptable, and in this regard cognitive neuroscience research seems to be ill-founded.

Finally, it looks as if contemporary cognitive neuroscience research does not on the whole meet the materialist challenge that it sets for itself of explaining the attribution of psychological properties to the nervous system by means of neurobiological ones. It fails, in particular, for lack of a foundational principle of naturalization that explains how the instantiation of a property of a psychological kind in the nervous system can be explained by the instantiation of a property of a neurobiological kind. Indeed, to the extent that it mentions such a principle at all, contemporary cognitive neuroscience shows a variety of materialist commitments, which all remain essentially dogmatic.

The examination of empirical work has also thereby revealed the absence of a strong bias in favor of a reductionist solution. Which is surprising in view of the fact that, in the last decades, reductionist claims have hinged crucially on the empirical discovery of nomic correlations between psychological and neurobiological properties, and that, on the other hand, the development of cognitive neuroscience has been essentially a history of establishing such correlations.

I am very grateful to Gregg Solomon for his penetrating remarks on an earlier version of this paper, which were of much help for putting it into final form.

Jean-Michel Roy
ENS-LSH (Lyon) and CREA (Paris)
France

NOTES

[1] Cf. Giere 1992, p XVI.
[2] Which I see as one aspect of the process of answering Q3; cf. *infra*.
[3] Indeed, it is important to note that foundational commitments of empirically developed theories can be variable in strength, as well as more or less explicit.
[4] Cf. Changeux, 1983.
[5] Cf. Patricia Churchland, 1986.
[6] "The scientific field of cognitive neuroscience, Gazzaniga writes in the opening lines of the first chapter, received its name in the late 1970's in the back seat of a New York taxi. One of us was riding with the great cognitive psychologist George Miller on the way to a dinner meeting at the Algonquin Hotel. The dinner was being held for scientists from Rockefeller University and Cornell University, who were joining forces to study how the brain enables the mind, a subject in need of a name. Out of that taxi ride came the term cognitive neuroscience, which took hold in the scientific community" (Gazzaniga 1998, p 5.)
[7] Cf. Kosslyn 1992.
[8] Cf. Imbert 1992.
[9] Cf. Rock 1974.
[10] Cf. Eimas 1989.

[11] Cf. Berthoz 1997.
[12] Cf. Posner 1982.
[13] For further analysis, see Roy 2000, 2001
[14] In the tutorial of the Conference in which this paper originates, however, a substantial amount of attention was paid to this confrontational aspect.
[15] In a psycho-physical parallelism perspective, for instance, where sheer correlation is substituted for ontological dependence, it can still be said that the attribution of properties P^1, P^2.... is true only if the attribution of properties N^1, N^2... is true.
[16] Cf. Nagel 1960.
[17] Cf. Hempel 1974.
[18] Cf. Mountcastle 1976.
[19] Cf. Crick 1990.
[20] Cf. Gilbert 1999.
[21] Cf. for instance Varela 1996 and 1999.
[22] Cf. however Imbert 1992.

REFERENCES

Albright D. and Neveille J, *MIT Encyclopedia of the Cognitive Sciences*, Neurosciences, 1999.

Berthoz Alain, 1997, *Le sens du mouvement*, Odile Jacob.

Changeux Jean-Pierre, 1983, *L'homme neuronal*, Fayard.

Crick Francis, 1990, "Towards a neurobiological theory of consciousness", *Seminars in the neurosciences*, 2, 263-275.

Churchland Patricia, 1986, *Neurophilosophy, Toward a Unified Science of the Mind-Brain*, MIT Press.

Eimas P. et Galaburda A., 1989, eds., *Neurobiology of cognition*, Cognition, Special issue.

Gilbert Charles, 1999, "Neural plasticity", in Allbright et al., 1999.

Giere Ronald, 1992, *Cognitive Models of Science*, Minnesota Studies in the Philosophy of Science, XV, Minnesota University Press.

Gazzaniga Michael et al., 1998, *Cognitive neuroscience, The biology of the Mind*, Norton and co.

Hempel Carl, 1965, *Aspect of science*, The Free Press.

Imbert Michel,1992, "Neurosciences et sciences cognitives", in Daniel Andler, ed., *Introduction aux sciences cognitives*, Gallimard.

Kosslyn Stephen and Koenig Olivier, 1992, *Wet Mind: The new cognitive neuroscience approach*, Free Press.

Mountcastle Vernon, 1976, "The world around us: Neural command functions for selective attention", Neural Sciences Research Bulletin, 16,#2 Supp.

Nagel Ernest, 1974, *The Structure of Science*, Hackett.

Posner M, Pea R., Volpe B, 1982, "Cognitive neuroscience: developments towards a synthesis", in *Perspectives on mental representation*, Mehler J., E. Walker et M Garrett eds.

Rock Irvin, *The Perceptual World*, Freeman and Company, 1974.

Roy Jean-Michel,

2000, « Naturalización de la mente y autonomía de la explicación mentalista », in *Mentes reales, Ensayos sobre la naturalizacion de la mente*, ed. J Botero et al., Siglo del Hombre Editores, Bogota .

2001, « L'émergence des Neurosciences Cognitives », in *Le cerveau, quelle histoire ?, Cahiers Alfred Binet*, n° *537*.

Varela, Francisco,

1996, "Neurophenomenology: A Methodological Remedy for the hard Problem", *Journal of Consciousness Studies* 3: 330-50.

1999, "The Specious Present: A Neurophenomenology of Time Consciousness", in *Naturalizing Phenomenology: Issues in Contemporary Phenomenology and Cognitive Science*, Petitot et al. eds., Stanford University.

ERNEST SOSA

CONSCIOUSNESS AND SELF-KNOWLEDGE

I still remember the impact that Gilbert Ryle made on me as a young student in the early sixties. I had recently been studying close-to-the ground behaviorist psychology, while another class took us aloft to Whitehead's *Process and Reality*. So Ryle in the next semester came as a welcome relief. Here were questions I could both understand and care about, answered through a mode of thinking and expression that was great fun. In this regard, Plato comes to mind as an early precursor, Ryle's great philosophical hero. But it was exhilarating to find a contemporary writing in this fresh way on issues of such interest. It gave me a boost in my early days as a student of philosophy, and helped sustain and increase my interest in the subject, for which I remain grateful. On recently re-reading the book my take on it is colored by my years as a professional philosopher, but I still admire it even now. This surprises me, as Ryle is no longer much cited or heeded, his brilliance long now under the shadow of Wittgenstein. Still I am left once again in no doubt that Ryle saw deeply into main and lasting issues of analytic philosophy, which to a large extent he himself helped define and place on the agenda, especially issues about the mind and its workings, but also broader issues of metaphysics and related philosophy of language, as in his work on categories and on systematically misleading expressions.

Today we celebrate his most famous and influential work, his book *The Concept of Mind*, published fifty years ago.[1] In this book already we find some main themes of Ryle's philosophical career, several of which have had long-lasting influence: the attack on dualism, for main example, and the various prongs of that attack, such as the distinction between knowledge how and knowledge that. Today I would like to focus on one such theme still now a focus of intense interest: the status of our "self-knowledge," as it has come to be known.

Jesus M. Larrazabal and Luis A. Pérez Miranda (eds.), Language, Knowledge, and Representation: 135-145.

1. RYLE AGAINST PRIVILEGED ACCESS

As part of his rejection of Cartesianism, Ryle discounted the importance and distinctiveness of our knowledge of our own mental or psychological episodes, states, moods, traits, and so on. He argued that the difference between how we know our own minds and how we know other minds is vastly overrated. Mostly we know others in the same ways we know ourselves. We need to see patterns in our reactions and in our behavior generally, for example, in order to discern a character trait, or a mood, or a motive, and the difference between others and oneself may be just a matter of quantity of easily available data. In principle someone else could be in just as good a position as oneself to make such an attribution. And character flaws such as vanity are more easily discerned by others than by oneself.

Ryle is of course aware of the supposed difference between episodes and traits, however, and wants to press his case even against the Cartesian citadel, *episodic* consciousness. At the beginning of the famous central chapter on "Self Knowledge," the target is defined with unusual closeness and formality:

> It is often held ... (1) that a mind cannot help being constantly aware of all the supposed occupants of its private stage, and (2) that it can also deliberately scrutinise by a species of non-sensuous perception at least some of its own states and operations. Moreover both this constant awareness (generally called 'consciousness'), and this non-sensuous inner perception (generally called 'introspection') have been supposed to be exempt from error. A mind has a twofold Privileged Access to it own doings, which makes its self-knowledge superior in quality, as well as prior in genesis, to its grasp of other things. I may doubt the evidence of my senses but not the deliverances of consciousness or introspection. (154)

Ryle would of course reject the existence of any fundamentally mental episodes as pseudo-antics of the ghost in the machine. But he objects to the doctrine of Privileged Access in ways mostly independent of that ontological issue. However, his main distinctive arguments rely on serious misconception. His argument against part 1 of the doctrine of Privileged Access, for example, presupposes that among the "supposed occupants of the private stage" may be found the following: one's knowing a certain fact, one's having a certain motive, one's being irritated or excited, one's being awake, and one's dreaming. But the Cartesian defender of Privileged Access to our states of consciousness would hardly suppose that any of these is a conscious *episode* of the sort to which we are supposed to enjoy privileged access.

As for his other main argument against that first part of Privileged Access, it is vintage Ryle. He objects that the doctrine leads to a vicious regress. For if we are inferring a certain conclusion from a certain premise, our inference will or at least may be conscious, in which case we must be aware of it, but this awareness itself is then conscious, which requires that we be aware of it in turn. And so on, *ad infinitum*. Is this really a vicious regress? No, in fact it is no regress at all; the argument rests on a confusion. It fails to distinguish between an episode's being a conscious episode, on one hand, and our being conscious *of that episode* on the other. Episodes can be conscious without one's being conscious of them. With this distinction in mind, the Cartesian avoids Ryle's alleged regress.

What of part 2 of the doctrine, according to which one can deliberately scrutinise, by a species of non-sensuous perception, at least some of one's own mental states and operations? In ostensibly arguing against this part of the doctrine, however, a doctrine that he himself has defined, Ryle mostly objects to a very different view: namely, the view that there are no mental states or operations that we cannot know about through introspection. Against *this* view he argues in two ways. First he appeals again to a vicious regress:

> ...[It]...will be allowed that there is some limit to the number of possible synchronous acts of attention, and from this it follows that there must be some mental processes which are unintrospectible, namely those introspections which incorporate the maximum possible number of synchronous acts of attention. The question would then arise for the holders of the theory how these acts would be found occurring, since if this knowledge was not introspectively got, it would follow that a person's knowledge of his own mental processes could not always be based on introspection. But if this knowledge does not always rest on introspection, it is open to question whether it ever does. (165)

As for his second argument against introspection as a source of knowledge, it draws from Hume the observation that some states cannot be known through cool and attentive consideration, since they are states incompatible with such cool reflection, as in a case where one is panicked or furious.

Neither of these arguments can carry much weight, however, since the fact that knowledge does not always rest on introspection does not refute the claim that it sometimes does. This would be like arguing that since we do not always know something to be round by seeing it to be round, therefore we never know it to be round that way. So Ryle's reasoning does not really touch the second part of the doctrine of Privileged Access as he himself defines it, since that part of the doctrine claims only that self-knowledge *sometimes* rests on introspection.

What is more, it is not clear that Ryle's definition properly identifies the most interesting form of the doctrine of Privileged Access. Why saddle the doctrine with the view that introspection must be a *directed* activity that cannot be performed in the presence of panic or fury? Why not allow that in such cases introspection is no more subject to deliberate guidance than is the belief that one sees light when one opens one's eyes to the midday sun. Any plausible form of Privileged Access must allow that an excrutiating headache can be *unavoidably* noticed.

So Ryle's arguments against Privileged Access do not carry the day. The issue that he placed on our agenda still remains unfinished business. We still struggle to understand the epistemology of self-knowledge even now, with vigorous discussion continuing in the most recent issues of today's journals.

2. PRIVILEGED ACCESS

It has been thought that Wittgenstein disposed of the problem of Privileged Access through a doctrine of avowals according to which pain reports have an expressive function and purport to state no facts. This is subject to the "Geach point," however: that the language of supposed avowals seems semantically much richer that any purely expressive vocabulary such as 'Ouch!' Thus the declarative 'I am in pain' can figure in antecedents of conditionals, as in 'When I am in pain, I normally take some aspirin.' However, as Crispin Wright has argued[2],

> ...what the Geach point signals is merely the presence of truth-evaluable content. It is powerless to determine that the standard use of a locution is to *assert* such a content. And now the expressivist thesis about avowals can be merely that the typical use of such sentences is as expressions rather than assertions.... [And] the suggestion ... will be that the appearance of the *epistemic* superiority of the self which avowals convey is an illusion created by attempting to find a home for features of such utterances which they carry *qua* expressions in the context of the mistaken assumption that they are ordinary assertions. (36)

However, as Wright argues, this approach still remains incredible even so, because of its incredible implications. For example, it implies that someone under torture who gives no ordinary outward sign of his agonizing pain cannot know about it as well as the torturers who read the characteristic signs of pain through sophisticated instruments, acquiring thus information denied to the victim herself.

The approach is incredible, but it is not Wittgenstein's, or so Wright would have us believe. In his view, the cash value of Wittgenstein's proposal about self-knowledge is that Privileged Access is not a consequence of the nature of our relevant psychological states, but is involved constitutively in the conditions of identification of what a subject believes.

> According to the Default View, it ... is simply basic to the competent ascription of the attitudes that, in the absence of good reason to the contrary, one must accord correctness to what a subject is willing to avow, and limit one's ascriptions to her to those she is willing to avow. (41)

Wright notes that the impression of dogmatism conveyed by the Default View can be dispelled by viewing the relevant psychological states as constituting a field of facts to which the subject is sensitive or responsive by virtue of his situation and cognitive capacities. This is what underlies the powerful analogy between our access to such facts about ourselves and our access through perception to observable facts about our environment. We have privileged observational knowledge concerning the field of facts about our surroundings to which we are appropriately sensitive or responsive by dint of our situation and cognitive capacities, including our perceptual capacities.

However, if we lack such a framework for our own relevant psychological states, if these cannot be understood on the models of colors and shapes of environing objects, then it is hard to see how the Default View can amount to much more than just a dogmatic stance. At the core of the onion we would find a peel with the disappointing inscription "This language game is played."

I agree with Wright about this. My own conclusion is that we have not found in Wittgenstein an acceptable resolution of our issue. Neither expressivism of avowals nor the Default View seems an acceptable resolution.

Christopher Peacocke has focused on the reason provided by conscious states themselves for beliefs about our psychology. He draws on his broader view about the conditions for concept-possession, and in particular the conditions for the possession of the concept of belief. It is in terms of such conditions that one can understand how it is that the judgment that p so appropriately leads the subject to self-attribute belief that p upon consideration. The very possession of the concept of belief requires it. But is Peacocke's proposal any less dogmatic than the Default View? Don't we need some reason at least to believe that in self-ascribing beliefs we are not likely to go wrong? In a response to Peacocke, Michael Martin raises this worry[3]:

> [We] ... can see Peacocke's account as implicitly offering to delimit in part the range of ... beliefs about which we have authority: namely, those which we can ascribe on the basis of conscious thoughts. But ... this is too broad a basis for authoritative self-ascription .... A too familiar phenomenon in philosophical enquiry is the realization that the proposition that one felt that one had a real conviction was so, turns out to be not the proposition that one identified initially, but really another proposition, which is a close cousin. (116)

So we can be wrong in self-ascription of such beliefs. We shall return to this objection below.

Gareth Evans has written: "The crucial point is [this] ... : in making a self-ascription of belief, one's eyes are, so to speak, or occasionally literally, directed outwards—upon the world."[4] Richard Moran has said similarly that determining what one intends is sometimes tantamount to making up one's mind.[5] According to Martin, there is a moral we could draw from Evans's approach (and similarly from Moran's):

> When the subject has her eyes directed outwards, they are directed on the world, as it is for her. Now, we should not assume that all of a subject's conscious states form part of the world as it is for her, not even all of her conscious beliefs. So, we might say, what we expect is that a subject should at least have authority with respect to those mental states which comprise in part her point of view on the world. Peacocke himself endorses a close link here between consciousness and the subjective point of view when he claims that 'The requirement that the reason-giving state is one which is, or could become, conscious is intimately related to our conception of an agent as someone with a point of view, one whose rational actions make sense to the subject himself ... given that point of view" (p. 96)." (117)

Despite the insights contained in this approach, it cannot be a complete account, not even of the restricted field of states constitutive of our subjective point of view. There is a gaping deficiency in the account, which suggests a different approach, one that will not account for the authority of such beliefs by appeal to the fact that they are based on and rationalized by certain conscious states.

The gap concerns our knowledge that we do *not* believe such and such, and our knowledge that we do *not* intend such and such and our knowledge that we do *not* seems to see anything red or any triangle. In no such case is our knowledge to be arrived at by means of first order inquiry or deliberation of any sort; nor is our knowledge based on some conscious state that rationalizes it. On the contrary, it is precisely the *absence* of a relevant belief or intention or experience that makes it a case of knowledge. Moreover, our belief that the state is absent seems clearly *not* to be based on any conscious state to which it is appropriately responsive. The more general problem for the approach in terms of an intermediate conscious state is this: What fundamentally accounts for our privileged access to what is present in our consciousness is unlikely to differ dramatically from what accounts for our privileged access to what is absent therefrom. So, even if the account of what is present must at some stage pay heed to some relevantly distinguished conscious states that help rationalize our pertinent beliefs, even so there must be something deeper that enables us to see what is shared in common between that case of privileged access and the equally privileged access that we enjoy to what is absent from our consciousness. We need therefore a deeper and closer look.

3. INTERNALISM AND THE A POSTERIORI

Here is Leibniz in Chapter 9 of Book IV of his *New Essays on Human Understanding*:

> Our direct awareness of our own existence and of our thoughts provides us with the primary truths *a posteriori*, the primary truths of fact.... [They can be called "immediate"] because there is no mediation between the understanding and its objects.

Compare Russell in *Problems of Philosophy*, ch. V:

> We shall say that we have *acquaintance* with anything of which we are directly aware, without the intermediary of any process of inference or any knowledge of truths.

An ambiguity is shared here by Leibniz and Russell. One's consciousness can harbor experiences that go unnoticed; unnoticed altogether, or at least unnoticed as experiences with an intrinsic, experiential character that they nevertheless do have. Yet just as one automatically jumps one's jumps, smiles one's smiles, and dances one's dances, etc., so one experiences one's experiences. And since experiencing is a form of awareness, one is thus in one sense automatically aware of one's experiences, precisely in experiencing them, even of experiences that escape one's notice and of which one is hence *un*aware, in another sense. What is more, it is not only her smile that the Mona Lisa smiles; she smiles her specifically enigmatic smile. Similarly one experiences not just one's experiencing but one's experiencing in the specific ways in which one does experience.

Which kind of awareness do Leibniz and Russell intend: (a) noticing, intellectual awareness, whereby one occurrently believes or judges the thing noticed to be present, as characterized a certain way; or (b) experiential awareness, whereby one is "aware" directly of an experience of a certain specific sort simply in virtue of undergoing it?

That distinction—between (i) n(oticing)-awareness, and (ii) e(xperiencing)-awareness—is important to epistemology as follows. From the fact that one is e-aware of something it does not follow that one is n-aware of it. Plausibly, when one is n-aware of something this will then automatically itself be something one would be justified in believing. But the same is not true of e-awareness. And the question will still remain as to how it is that one can "notice" the things that one does notice in one's experience. More specifically the question will be this: To notice a fact about one's experience at a give time is to believe correctly that it is so, but just a guess will not be enough: it is required that the correct belief be also at a minimum justified, or reasonable, or epistemically appropriate, or some such. And this is just the question of what sort(s) of experience can be taken note of with epistemic justification merely through believing it (them) to be present within one's consciousness at the time.

Foundationalists through the ages have tried to understand how we can be justified foundationally in a certain belief. And in doing so they have appealed crucially to what is "given" in one's experience, or to what is "present" to one's consciousness. So they have appealed to what we are "directly" aware of. But this requires that we be clear on the kind of awareness that is here invoked: is it e-awareness or is it n-awareness. If the latter, then we seem *not* to have the desired explanation, since the concept of "noticing" seems itself epistemic in a way that unsuits it for the explanatory work that it is being asked to perform. What we want is an explanation in nonepistemic terms that will explain how it is that a noninferential, foundational, belief can acquire epistemic status in the first place, so that holding it is not just arbitrary, and so that inferences from it can lead to epistemic status in turn for the conclusions of these inferences. So our explanation cannot properly rest with "noticings" that are supposed to already have epistemic status. The question will remain as to how *these* beliefs constitutive of the "noticings" have acquired their epistemic status.

Thus are foundationalists driven to postulate some mental phenomena epistemically more primitive than any "noticings" or beliefs, which means that the conscious states that are "given in" or "present to" consciousness must be understood as cases of e-awareness: that is to say, as states that are constitutive of the subject's total mental life at the time, including both the constituents of that life-slice of which the subject is meta-aware and also those which escape her notice at that time. But we now have a problem for the epistemology of the given, the Problem of the Speckled Hen. The problem is one that worried Roderick Chisholm since his 1942 publication in *Mind* of a paper with the title "The Problem of the Speckled Hen." The problem is one that Chisholm found in Ayer's *Foundations of Empirical Knowledge*. The problem is essentially that of the gap between e-awareness and n-awareness, where much in the multiplex character of our experience can escape our notice, and can even be mistakenly characterized, as when one takes oneself to be able to tell at a glance how may speckles are in the image, and there are really ten large quite separate speckles in total, rather than eleven as one believes.

When he introduces the problem, by the way, Ayer acknowledges that he owes it to another philosopher. Interestingly for us, that philosopher was Gilbert Ryle.

The full dimensions of the problem posed for foundationalist epistemology have not yet been properly appreciated, or so it seems to me. If the foundationalist wishes to have a theory and not just a promisory note, he needs to tell us *which* sorts of features of our state of consciousness are the epistemically effective ones, the ones such that it is *by corresponding to them specifically* that our basic beliefs acquire epistemically foundational status. Having a visual image with forty-eight speckles seems not to qualify, whereas having a visual image with three speckles may (at least when the speckles are large and separate enough). What is the relevant difference here?

Our distinction between e-awareness and n-awareness also reveals a further limitation of Ryle's definition of Privileged Access, whose first clause reads again as follows:

(1) that a mind cannot help being constantly aware of all the supposed occupants of its private stage.

The question is now as to the kind of awareness involved in this clause: is it e-awareness or is it n-awareness. If the former, then the clause is trivially empty, amounting only to the claim that a mind cannot help undergoing whatever experiences it undergoes. If the latter, then the claim is absurdly strong, and not something that any sensible form of Privileged Access can require: no-one sensible can require that one must actually notice every aspect, nor even every constitutive or internal or intrinsic aspect, of every experience one undergoes. Our visual experience, for example, surely is multiplex enough that much of its intricacy escapes our notice at any given time.

So the question naturally arises as to how it might be advisable for Privileged Access to restrict its claim (1) within more defensible bounds. What distinguishes the occupants of one's consciousness that one can properly know to be present "by direct inspection"? Or, to ask the same question another way: What is the kind of state which by being given in or present to one's consciousness becomes a source of foundational status for a corresponding belief? [6]

4. KNOWLEDGE BY ACQUAINTANCE

Foundationalists have offered various accounts of how it is that one can derive foundationally justified status for one's beliefs by focusing on the character of one's own conscious experience at the time. It seems to me that none of these attempts is successful. But in the time available, rather than discuss why earlier attempts fail, I would like to offer instead a positive view that seems to me more likely to succeed when properly developed and defended.

What distinguishes the case of 48 speckles where one guesses right and does not know from the case of 3 speckles where one does enjoy foundational knowledge by acquaintance? The relevant distinction is that in the latter case, unlike the former, one's belief is both *safe* and *virtuous*, or so I wish to suggest. It is safe because in the circumstances not easily *would* one believe as one does without being right. It is virtuous because one's belief derives from a way of forming beliefs that is an intellectual virtue, one that in our normal situation for forming such beliefs would tend strongly enough to give us beliefs that are safe.

One does not know foundationally that one's image contains 48 speckles even if one's image *does* in fact containt 48 speckles, and one's belief hence corresponds precisely to what is then given in one's consciousness. One fails to know in that case because too easily one might have believed that one's image had 48 speckles while it had one more speckle or one less. But that is not so for the belief that one's image has 3 speckles.

But it is not sufficient that one's belief be thus *safe*. Consider the belief that 4,567,888 + 3,767,923 = 8,335,811. Not easily would anyone believe this without being right, since not possibly would anyone believe it without being right. However, if the way one arrives at one's belief is by a method according to which the sum of any two seven-digit numbers is any seven-digit number, then one fails to know, despite one's belief being perfectly "safe," in the sense defined.

That is why knowledge requires one's belief to be not only safe but also virtuous. But the relevant notion of an intellectual virtue needs to be explored more fully, as it remains obscure. I conclude with a sketch of some relevant considerations.

5. KNOWLEDGE AND VIRTUE

Knowledge requires nonaccidentally true belief; this casts some light on what knowledge is, but light dimmed by the imprecision in our notion of an accident, which itself needs explaining in the final analysis. More adequate is a notion of dependent mirroring. Compare the belief that this is a ripe tomato, believed of a ripe tomato amidst a bushelful of wax replicas. Arguably, one's belief then *does* mirror the truth, dependently on the circumstances. But it is still in some relevant sense or respect true only by accident. Each of the items in that top layer is believed by me to be a ripe tomato, but I am right only in this one case, which my perspective does not distinguish in any relevant respect.

Suppose I fancy myself a connossieur of tomato ripeness, but suffer from a rare form of color blindness that precludes my discerning nearly any shade of red except that displayed by this particular tomato. Therefore my judgments of tomato ripeness are in general apt to be right with no better than even chance. But when it's the particular (and rare) shade of red now displayed, then I am nearly infallible. Oblivious to my affliction, I issue judgments of tomato ripeness with abandon over a wide spectrum of shades of red. Assuming that, unknown to me, the variety of tomato involved always ripens with this shade of red, my belief that this tomato is ripe *is* in step with the truth. Nevertheless, in some relevant sense or respect I am right only by accident. The problem is that I am not adequately sensitive to the presence of the conditions on which my mirroring of ripeness is dependent. I still issue my verdicts of ripeness even when those conditions are absent. In this respect I am like Magoo in his perceptual judgments, which he does not restrict to the very narrow bounds within which they are reliable.

Compare a basketball B held by S at the top of an incline I. It is true, let us suppose, that (C) if B were released by S at t, it would roll down I. And this is true for reasons of two sorts: because of factors pertaining to the internal or intrinsic make-up of B at t: e.g., that it is rigidly round at the time; and because of factors external or extrinsic to B: e.g., because it is not glued to the top of the incline, because it is in a gravitational field, etc. Moreover, the truth of the conditional might be dependent on factors of two sorts: the external and the internal. While cradled in S's hands, B's internal pressure may hover at the limit above which, if released by S it would explode upon contact with the incline, rather than rolling down. So it may be true that if it were released by S at t it would roll dependently on its condition at t, since its being in just that internal condition is too precarious a matter for it to be the case outright that it would roll if released. Alternatively, incline I may be in a moment of calm at t, even though at t-e and at t+e it is swept by winds that would blow B away ratherthan allowing it to roll down. So conditions of two sorts may be involved in the truth of a conditional such as C, in such a way that C is true not outright but only dependently on the holding of the conditions; some such conditions are "internal," while others are "external."

Virtues are internal conditions that enable an entity to attain desirable ends (of the entity or of its users) relative to certain circumstances, conditions in virtue of which it is true that the entity *would* succeed in those circumstances. So a condition may be a virtue relative to one end-circumstance pair <E, C> while it is not a virtue relative to another <E', C'>. Of course, the context of thought or discussion may set the relevant ends and circumstances well enough that they need not be mentioned specifically, in which case we may and do speak of virtues without qualification. But there is always a danger of ambiguity, especially when we range in thought or speech through the realm of the possible: do we then retain the relativization to our actual ends and circumstances, or do we shift to our ends and circumstances in the supposed possible situation? This distinction is a source of ambiguity.

There are virtues of two sorts: those included in one's fundamental nature, constituted by one's innate capacities and aptitudes; and those part of one's "second nature," one's character pliable at least to some degree, at some remove. Of these, some may derive from blind habituation, from one's earliest upbringing perhaps, subject to very limited rational control. Nevertheless, at least in normal human development, there is much in one's character, moral and intellectual, due to one's own rational control. The acts, practical or theoretical, that derive from such components of one's character, from one's rational second nature, are under one's motivational control, at least to some extent, at some remove. These acts reflect on one's character, moral or intellectual, and specifically on those components of one's character for which one is to some extent, at some remove, responsible. With regard to such acts and the character they reveal, one is subject to suasion, and thus a proper participant in rational deliberative dialogue, and properly subject to praise and blame.

Ernest Sosa
Brown University and Rutgers University
U.S.A.

NOTES

[1] *The Concept of Mind* (Chicago: University of Chicago Press, 1949); relevant parenthetical references in the text will be to this. (And I shall follow the same pattern, mutatis mutandis, for other texts to be cited below.)
[2] In his “Self-Knowledge: the Wittgensteinian Legacy,” ch. 1 of *Knowing Our Own Minds*, ed. by C, Wright, B.C, Smith, and C. Macdonald (Oxford: Oxford University Press, 1998.
[3] In his “An Eye Directed Outward,” ch 4 of *Knowing Our Own Minds*.
[4] Gareth Evans, *The Varieties of Reference*, ed. John McDowell (Oxford: Oxford University Press, 1982), p. 225.
[5] Richard Moran, “Making up your Mind: Self-Interpretation and Self-Constitution,” *Ratio* NS I (1988): 135-51.
[6] The prospects are quite dim for a solution to the Speckled Hen Problem within the confines classical foundationalism, or so argues my “Privileged Access,” in *Consciousness: New Philosophical Essays*, edited by Quentin Smith and Aleksander Jokic. Oxford: Clarendon Press, forthcoming.

RONALD DE SOUSA

TWELVE VARIETIES OF SUBJECTIVITY: DIVIDING IN HOPES OF CONQUEST

Subjectivity. *See consciousness* (Chalmers 1996, index, 413)

Subjectivity is a theme common to many of those philosophers eager to deflate the ambitions of cognitive science. The claim is that persons differ from all other things in that they cannot be exhaustively described in the third person. Any attempt to do so will fail to capture something about every human being that is essentially subjective. This expression covers many things, and the word sounds all the more impressive for the fact that the things it purportedly designates are lumped into a very mixed bag. When lumped together as if they constituted one hugely complex problem, they tend to induce a sense of hopelessness. Which is exactly what some of the champions of subjectivity count on to preserve its mystery and irreducibility.

Among the champions in question, some of the most famous are Taylor (1989), Nagel (1986), and Searle (1992, 1997)).

Here is sampling of some of their claims. First, Nagel:

> "[T]he purely objective conception will leave something out [viz., the subjective content of "I am Thomas Nagel"] which is both *true* and remarkable" (Nagel 1986, 64).

Next, Charles Taylor:

> There are certain things which are generally held true of objects of scientific study which don't hold of the self:....
>
> 1. The object of study is to be taken "absolutely", that is, not in its meaning for us or any other subject,... ("objectively").
>
> 2. The object is what it is independent of any descriptions or interpretations offered by any subjects

Jesus M. Larrazabal and Luis A. Pérez Miranda (eds.), Language, Knowledge, and Representation: 147-164.

3. The object can in principle be captured in explicit description;

4. The object can in principle be described without reference to its surroundings. (Taylor 1989, 33-34).

Next, Searle:

> Conscious mental states and processes have a special feature not possessed by other natural phenomena, namely subjectivity.....[M]uch of the bankruptcy of most work in the philosophy of mind and a great deal of the sterility of academic psychoanalysis over the past fifty years,... have come from a persistent failure to recognize and come to terms with the fact that the ontology of the mental is an irreducibly first-person ontology. (Searle 1992, 93,95).

> Consider, for example, the statement 'I now have a pain in my lower back.' That statement is completely objective in the sense that it is made true by the existence of an objective fact.... However, the phenomenon itself, the actual pain, itself, has a subjective mode of existence. (Searle 1992, 94).

To these claims about the irreducibility of subjectivity, two forms of resistance are possible. One is to claim that the multifarious problems posed by consciousness and subjectivity actually all reduce to one. That one may or may not be currently soluble, but at least one has one problem and not many. That strategy is adverted to (though not adopted) in a recent book by Lycan, who defends "a weak version of Brentano's doctrine that the mental and the intentional are one and the same.... It would follow that once representation itself is (eventually) understood, ... I do not think there will be any 'problem of consciousness' left" (Lycan 1996, 11)[1]. A similar strategy may be implied in the recent books by (Tye 1995) and (Dretske 1995) defending a representationalist theory of consciousness.

The line I propose to pursue here is the opposite. It starts from the consideration that big mysteries are sometimes made of a lot of little tricks, and so might yield to a divide-and-conquer strategy. I suspect this is true of the mysteries of consciousness: if the "problem of consciousness" is not one, but many, and if each one can be successfully dismissed or solved along naturalistic lines, then by this different route we shall reach the same goal, of bringing it about that there not be any 'problem of consciousness' left.

I do not aim to demonstrate this large claim here. I concentrate only on the term 'subjectivity', and propose merely to make a start on the first phase, consisting in drawing up a list of ostensibly different problems of subjectivity. If some should turn out to be reducible to others, so much the better. But if not, then each variety of subjectivity might be tackled singly, and this might indeed contribute to a "natural history" of the human mind, in such a way as to bring it all under the aegis of science.

In his perceptive comments on the version of the present paper as presented to the ISCC conference, Jean-Michel Roy urged that by concentrating on the diversity of claims using the word 'subjective' made by philosophers, I risked missing the point which only a proper conceptual or phenomenological analysis could reveal. But the concept lives in what people use it to mean. No conceptual analysis, therefore, can avoid taking into account what those who use the concept have used it to do. Phenomenological analysis suffers from similar problems[2]. Still, one might ask what the root word 'subject' itself suggests as to what the core of subjectivity might mean. Two sufficient conditions then suggest themselves: either that we are talking about items to which only the subject has epistemic access, or that we are talking about items that are ontologically distinct in somehow pertaining only to the subject, as claimed by Searle in the passage just cited. The former can happily be conceded by any materialist. The latter begs the essential question of whether it makes sense to speak of an ontological category which is essentially defined in terms of conditions on epistemic access to it. But these features do not exhaust the claims made for subjectivity and its consequences for our understanding of the mental. That is my justification for undertaking the sort of botanizing I propose in what follows.

To give the flavour of the strategy to which this botanizing is supposed to contribute, here is an example of how confusion between various senses of subjectivity can be misleading.

Berkeley argued against the distinction between primary and secondary qualities that are all equally "ideas existing only in the mind": (Berkeley 1957, 27, 30). The subjectivity of secondary qualities, in the sense of their relativity to the observer's mind, can be shown to attach equally to primary qualities. If we resist the idealist conclusion, we can re-interpret this remark as implying that the perception of all qualities depends on the interaction between the external world and states of the subject's sense-organs. This assumes that if the appearances of things are *relative* to the sensory and conceptual apparatus of the perceiver, this entails that their attribution to the outside world is mere *projection*, with no objective correlates beyond themselves. The argument conflates *phenomenology* — the quality of experience — *relativity* to an observer, and *projection* — the attribution of a property to the outside world which is actually entirely resident in or manufactured by the observer. This conflation is plausible in the extreme case in which some quality attributed by an observer to a target depends totally on the perceiver and not at all on the target. For there is then nothing to the property in question except the observer's experience of it, and relativity collapses into projection. But no lesser degree of relativity can effect this collapse. At most, Berkeley's arguments show that the conflation of these different senses of subjectivity leads to idealism. This is not what the modern champions of subjectivity intend, but it may turn out to be the logical consequence of their strategy nevertheless. To some of us, this is reason enough for avoiding the conflation.

Phenomenology, relativity and projection are only three of the possibly distinct senses of subjectivity that have been adduced against materialism. In what follows I distinguish twelve basic varieties — senses, readings, interpretations, or aspects — of subjectivity or 'the subjective'. Some, as we shall see, might easily be further divided. Moreover, I am not confident that they are exhaustive. But I remain unconvinced that any form of "irreducible subjectivity" presents an obstacle to physicalism, and I offer the hope that tackling each variety singly will may make it easier to pre-empt their use as a medieval mace to whack wicked reductionists over the head with. Here then, is my list.

1. PERSPECTIVE

An individual is *somewhere in space-time*, and not somewhere else. Except for God, of course, who was invented to instantiate all contradictions in blessed harmony. He's everywhere and everywhen, though at the same time, as it were, not in time or space[3]. But the upshot of this is that every individual has a point of view, a perspective, and apprehends the world, so far as it can apprehend the world, from somewhere and not nowhere[4] (Nagel 1986). If taken in isolation, the feature of being somewhere in particular affects all kinds of individuals, not just humans. But only those individuals that can *view* something can presumably have a *point of view*. Thus Searle again:

> Subjectivity has the further consequence that all of my conscious forms of intentionality that give me information about the world independent of myself are always from a special point of view. The world itself has no point of view, but my access to the world through my conscious states is always perspectival. (ibid. 95).[5]

In itself, however, that could be true of any other living thing. Nor is it a requirement to be alive: an artificial eye has a point of view. More generally, as shown in the excellent discussion of this subject in (Proust 1997), aspectuality can be seen as a consequence of mere differences of informational channels, and doesn't therefore require any level of consciousness.

Perspective might itself be of two kinds. This can be seen by asking: Does a *still* camera have a genuine *point of view*? One reason to deny this is that for a still camera there is nothing that corresponds to the difference between locality in time, and locality in space. For a living individual, these pose slightly different problems. For there are different ways in which we might care about the effects of our actions in distant space, and in different times. Time is asymmetrical in this sense (among others): we care more, or quite differently, about what happens in the future than about what happened in the past. But although the things we care about may, of course, be unevenly distributed, space has no uniformly privileged direction. So temporal perspectivity appears seems to constitute a more serious species of subjectivity than the spatial kind.

Now perspectivity is sometimes equated with subjectivity in general, as suggested in the last quotation from Searle above. Yet subjectivity is also associated with the self, and the temporal form of perspectivity actually causes problems for the view that my *self is* my subjectivity. This is because changes in perspective, especially in temporal perspective, change the relative value of different prospects. For example, as (Ainslie 1992) has pointed out, we seem to discount the future at a hyperbolic rate, so that the closer prospect can surpass the more distant in apparent value, rather as a low building can loom higher than a tall one when one is up close to the former. Where such changes occur, which perspective is the right one, that is, truly *mine*? Are there as many individual selves as there are perspectives? In a recent article, Galen Strawson answers in the affirmative: each of us is many brief, material, successive selves, he says, strung like pearls on a string (Strawson 1997). Before him, Derek Parfit (1971, 1984), is famous for advocating a similar view.

Suppose I get my friends solemnly to promise to put me gently to death when I become gaga, because I would rather die than be gaga. What if, once I become gaga, my priorities change? Now I don't want to die: I would rather live and be gaga. Do my friends still "owe" me euthanasia, against my present wishes? Actually, the answer is always No, but for a different reason. If I'm a different self from what I was when they made their promise, then you can't be bound by *him* (*i.e. me-then)* to do anything to or for me (i.e. *me-now*). But if I'm the same person, then I can now relieve you of your obligation to me if I change my mind. The facts about perspective, then, appear to be neutral in practice between the Parfitean and a traditional concept of the self, but they seem to be significantly different in theory.

Note, however, that in articulating the problem of the asymmetry of time we have to introduce an additional factor: what's involved is not just being at a certain place and time, but envisaging what is seen from that point of view as affording possibilities for *agency*. Make that our second form or aspect of subjectivity.

2. AGENCY

Agency is presumably not an aspect of human individual subjectivity that we concede to inanimate individuals. As a human, one experiences oneself as having the power to choose and act. The fact of being an agent, as has been often stressed,[7] is a form of subjectivity in the precise sense that I, the subject, and I alone can decide what I will do, although — depending on your own particular stance on the tricky antinomies of free-will — all sorts of circumstances can determine what I in fact end up deciding. Whether I am free to decide to do or not to do A, compatibly with your ability to predict which I will do, is a conundrum that I won't discuss. I'll only assert the obvious, namely that whether or not you can predict what I will do does not change the fact that I do in fact experience myself as *deciding*.

Perhaps this apparent fact about the irreducibility of decision is really just an effect of perspective. From my own point of view as an agent, I can't take my reason for action to *causally-determine* my action without failing to decide; but failing to decide is just another decision. (Compare: I never can directly see my own face: would I be right in concluding that my face is different from everyone else's in some crucial way that makes it invisible?)

The first locus of the claim that agency is a form of subjectivity is probably Descartes[8], though he didn't say it in so many words. But it takes just a little teasing out to get from the claim that the will is infinite to the present thesis. The infinitude of the will is unfortunately compatible with complete powerlessness. So the measure of the will's freedom has nothing to do with its effectiveness in bringing about any change in the world. Furthermore, this infinite freedom says nothing about the origins of our desires. An admittedly simplistic argument suggests that the infinitude of the will's freedom is also compatible with there being absolutely none of my desires that originates in myself. For whatever my desire, I cannot deny that it might have been different had I different genes or had I had a different life. In other words, my desire must have come from causes ultimately traceable to my genes and to my environment. But since I am not in any sense the author of either my genes or my environment, it seems to follow that I'm not the author of my own desires either. Whatever one may think of this somewhat fishy argument, it remains true that the "freedom of the will," which I'm equating with the subjectivity of agency, cannot be denied: whenever I am made conscious of a set of possible choices, choosing is not so much something I can do, as something I cannot forbear to do, regardless of the origins of my grounds for making it or of whether my choice makes any difference to what results.

3. TITULARITY OR OWNNESS

One of the specific ways in which my power of agency is "essentially subjective" is that my actions are *mine* in a peculiar sense of the word. Sergio Moravia (1995) has labeled "titularity" the fact that my mental attributes (including but not only including qualia) are my own in a unique sense of ownership. This sense of ownership is indeed peculiar. It is different from the sense in which I own my bicycle, different from the sense in which I own my hair; different from the sense in which, on some views, I own myself and no other person can logically own me; and different again from the sense in which I "own myself" and no one else can (ethically) own me. For if it is unethical for some person to own another, then it is not metaphysically impossible. But it would seem to be not merely unethical, but metaphysically (or logically) impossible for the slave-owner to own his slave's experiences.[9] The point has been made by Tye (1995,

10-11,71ff), who distinguishes two problems raised for materialism by this feature. We might call these the two *conditions of special ownership*. One is that every mental state necessarily belongs to someone or other; and the second is that every mental state necessarily belongs to whoever it belongs to and not anyone else.

This is a particularly good example of the *mystifying* function of these declarations of subjectivity. For the special sense of ownership involved here is not, in fact, exclusive to mental states, but belongs to a large class of predicates. It was described long ago by Aristotle, in connection with what commentators have named "dependent particulars". There are two senses in which we can talk about the *whiteness of this paper*: one refers to a specific shade of white, and in that sense the whiteness of this paper might also belong to, or characterize, some other surface. But in another sense it is logically impossible that the whiteness of this paper should belong to anything else. We can reidentify this paper, even if it has changed colour, but there is no way that we can reidentify its whiteness independently of it. The paper exists independently of its whiteness, but not vice versa.[10] Tye points out that non-mental actions, such as one person's laughter, or her walk, also meet both the conditions of special ownership. Events, even those involving no agency at all, exhibit the same feature. My pen's falling to the floor is not something that logically could pertain to nothing, nor is it something that could pertain to anything other than my pen.

Besides titularity, the legal notion of ownership involves two features that have figured prominently in characterizations of what it means for something to belong to me. One is that I have a special right to use it: I have, as the phrase goes, *priviledged access* to it. The other is that I have the right to exclude others from my property. You might call this the right of *privacy*, and where it concerns my beliefs about myself, it amounts to their incorrigibility by any one else. These two, then, constitute the next two forms of subjectivity. They are commonly confused, at least in the terms used to describe them. But if we keep in our minds the difference between access and exclusion, it seems plain that they are indeed separable doctrines. I might enforce my right of access to my property, while not excluding anybody else. The converse seems to make less psychological sense, but is not logically impossible.

4. PRIVILEGED ACCESS

It was long a dogma of the philosophy of mind that one of the defining characteristics of mental states was their privacy, that is, their inaccessibility to other observers. Tye sees this as one of the aspects of ownership: "My pains, for example, are necessarily private to me. You could not feel any of my pains." (Tye 1990, 71). But clearly privacy is distinct from that other feature of ownership, privileged access. This may well be one of the features that the champions of subjectivity have in mind, but nowadays the issue of access is not generally regarded as clear in either direction. This is partly due, no doubt, to the influence of Wittgenstein's attack on private languages. On the one hand, we have gotten used to talking about mental states which are clearly enough my own, but to which I have no access either because they are repressed in the "Freudian Unconscious" or because they pertain to the "Helmholzian Unconscious" (Johnson-Laird 1988, 354). Conversely, in the light of some recent thought experiments, the impossibility of accessing another person's mental states can no longer be asserted without begging just the sorts of question at issue in debates about materialism.[11]

5. THE INCORRIGIBILITY OF APPEARANCE

One of the political privileges of privacy, in sense in which we speak of a right to privacy, is the right to keep others out. Under the last rubric I have focused on the subject's access (which turns out to be dubious). What then of the subject's converse right to exclude others?

Objective reality is more than meets the eye. No one subject, it seems, is ever in a position to exclude others from all facets of Reality. "Mere" appearances, as we call them, on the other hand, are subjective. We inherit from Plato one of the reasons for drawing this contrast: appearances change, while reality supposedly stays the same. But this is quite wrong-headed. If you were aware of an "appearance" which never changed, it would be a pretty sure sign that you were having a hallucination. Part of the changes brought to appearances are due to perspective, which I've already talked about; and if you didn't see something in perspective and from your own point of view that would prove that it was not objective. If I thought I saw a circle from the side and it appeared circular, then I'd have to conclude it wasn't really a circle. The supposed subjectivity of appearance, then, lies in its *incorrigibility*. It is only I that am incorrigible about what appears to me; anyone else fails to have equal authority with mine. Conversely, what *seems* to be the case is the only thing on which (for example) Descartes allows that I am incorrigible (Descartes 1984-1985, 29). Incorrigibility thus emerges as a form of subjectivity independent of those already listed, because it is logically possible that propositions concerning perspective, agency, ownership, and even privacy (and also qualitative experience and seeing-as, which we shall get to in a moment) might all be corrigible on the basis of objective evidence. Incorrigibility has a low status these days: most philosophers agree that if any candidates presented itself it would turn out to be an illusion (Gopnik 1993). An inverse relationship holds between the empirical content of a claim and the degree of its certainty. It is a plausible principle, even if we do not cling to strict verificationist or falsificationist dogmas, that there is a direct correlation between corrigibility and content. (Sellars 1963)

6. PROPRIOCEPTIVE SENSE

Among the things I own in some peculiar sense, though not in the peculiar sense just discussed, is my own body. Herein lies one more trademark of subjectivity. In the most common case, the proprioceptive "sense" designates the awareness we have of the position of one's limbs. Try this: close your eyes and touch your nose with your index finger. You may miss, but not by much. Ramachandran and Hirstein have described a delightful experiment in which I can actually find the tip of my nose to be displaced to where the tip of your nose actually is actually displaced:

> [T]he subject sits in a chair blindfolded, with an accomplice sitting at his right side.... facing in the same direction. The experimenter then stands near the subject, and with his left hand takes hold of the *subject's* left index finger and uses it to repeatedly and randomly tap and stroke the nose of the accomplice, while at the same time, using his right hand, he taps and strokes the subject's nose in precisely the same manner, and in perfect synchrony. After a few seconds of this procedure, the subject develops the uncanny illusion that his nose has either been dislocated, or has been stretched out several feet....." (Ramachandran and Hirstein 1997, 452).

What I find particularly intriguing about this illusion, is that in fact this "sense" which guided your hand is not a sense at all, insofar as it has no "organ". What's more, it is clearly a form of subjectivity, insofar as only the *subject* can make the relevant observation. We can't *have* someone else's phantom nose illusion. But it's not just a quale or bundle of qualia.

Here again one might divide even more finely. For the special proprioceptive consciousness of one's own face seems to form a distinct class by itself. It is not so easily explained as the nose-displacement illusion, and even more interesting, Jonathan Cole has described severe disturbances in self-concept and interaction with others in patients suffering from "Möbius syndrome", which involves an inability to move any of the muscles of facial expression (Cole 1998). This inability is described as inhibiting the development of a sense of self, no doubt largely because it makes impossible facial imitation which, from the earliest days of a baby's life, establishes one's sense of one's own emotions in some sort of concert with the emotions of others. Cole cites (Meltzoff and Gopnik 1993)'s observation of imitation in infants as suggesting "that in early experience babies learn something of emotion, and how it is experienced, by taking the facial expressions of others and, by imitation, feeling their own faces to be like others" (Cole 1997, 481). More of this under a later heading (see §9: "the subjectivity in intersubjectivity"). At this point I wish only to point out that no such mechanism could make sense unless there existed the sort of pseudo-sixth sense that is proprioceptive perception of one's own face.

But is this a problem comforting to the mysterians? No. On the contrary, all these proprioceptive phenomena, both common and exotic, are highly suggestive about the physical, neurological origins that are likely to give rise to them.

7. IPSEITY

When I refer to myself, I am not just referring to the person who happens to be me. This is a point developed in a number of papers by H.-N. Castaneda (e.g. Castaneda 1988) and by John Perry (1979). The latter's vivid example has him noticing a trail of sugar in the supermarket. He identifies its source as someone whose cart contains a leaking bag of sugar, who is unaware of it, and who has apparently been all over the supermarket. But for a long time fails to identify the person thus "identified" with *himself.*[12]

Is knowing that *I (Ronnie) am I* a real piece of knowledge? God couldn't know it, though he could know that the writer of the previous sentence is Ronnie, or any number of other statements identifying me with myself under two different descriptions.

A similar point could be made about perspective: since God is everywhere, he necessarily lacks perspective. Is this a limitation on God's supposed omniscience? Whatever the answer, it is tempting to think that ipseity is merely a side-effect of perspective. Tempting, but wrong: for the facts of perspective are entailed by the existence of spatio-temporal particulars in space-time; not so ipseity, since it would be theoretically possible for all information I have about myself to be devoid of perspective, and for all my desires to be formulated in entirely general terms. Sober and Wilson have suggested that ipseity is an adaptive trait which allows a self-interested individual to channel benefits to itself without having to burden itself with large amounts of discriminatory information. "This speculation," they add, "entails a small irony. People use the concept of "I" to formulate the thought that they are unique. Yet, part of the reason that people have this concept is that they are *not* unique...." (Sober and Wilson 1998, 214, 350). Sober and Wilson also correctly point out that ipseity (which they call "self-recognition") differs from "self-awareness" in that "self-recognition does not require that the individual be a psychologist", i.e. think of themselves as having beliefs and desires (Sober and Wilson 1998, 216).

8. TONE OR COLOUR

What is it like to be *you*? It's not obvious either that Descartes was right about the transparency of that consciousness, nor that there isn't anything it's like to be me, nor that it's somehow reducible to all the others, or to some subset of qualia. An individual *tone, or colour* is thus subjective in what seems to be yet another irreducible sense. Nevertheless, the colour of my life may supervene on many physical properties, just as the colour of a surface supervenes on a number of properties of texture, light, and relational properties computed by our visual system in ways determined by complex ecological factors (Thompson 1995). What is distinct about this form of subjectivity is that it concerns not sensory experience in general, but one's experience of oneself in particular. It is precisely not reducible to ipseity, however, if the contrast I just borrowed from Sober and Wilson between ipseity and uniqueness is a real one. My feeling-of-being-me may well be different from anyone else's analogous feeling, and indeed is likely to be so insofar as it supervenes on a number of factors that determine different aspects of our experience of ourselves.

9. THE SUBJECTIVITY IN INTERSUBJECTIVITY

My identity is, in part, intersubjective. I mean by this that it is causally constituted by my being able to gauge the state of my own mind, and particularly my own emotions, in interaction with others. I note three aspects of this interaction. First, grown-ups tell children what they feel, more or less effectively, resulting in adults who know more or less what they feel. I am not sure quite how to analyse the capacity to be so trained to recognize one's own emotions; obviously it presupposes that there must be something that one is being trained to recognize, but it doesn't follow that there must already be full-fledged emotions awaiting recognition. For it may be, as Sue Campbell has stressed, that expressing emotions contributes crucially to the *determination* of the emotion, both in the sense of bringing the emotion into being and in the sense of making it to be just this emotion and not another (Campbell 1998). Again, there would be no such possibility were there not some subjective side to the interpersonal transaction which is an expression of emotion. One creates one's subjective sense of one's own emotions by comparing them to others. But this suggests something of a paradox, since *subjectivity seems to be a precondition, or presupposition, of its own cause*. For there surely can't be inter-subjective engagement if there are no subjectivities between which the engagement takes place. The solution to this puzzle no doubt lies in a developmental and dynamic perspective, allowing us to see how the intersubjective and the subjective appreciation of one's emotional self develop together. One piece of the puzzle may lie in this third observation: Meltzoff and Gopnik (1993) have shown that infants are able to imitate facial expressions in the very earliest days of their lives, at a stage when one would be disinclined to ascribe to them anything like a sense of self. Nevertheless it would be surprising if this capacity didn't play a role in the acquisition of emotional expressions and through them of a sense of self. Jonathan Cole puts it this way:

> Through imitation a face can be assimilated from visual experience, through proprioception, into felt experience: something can be taken from being "out there" in another, to being "in me." (Cole 1998, 110)

If so, then certain forms of interaction would lie at the base of the acquisition of subjectivity in some of the other senses I have sketched.

10. PROJECTION

Though most mentions of subjectivity are intended to prove our species superiority, it does sometimes happen that the connotation of subjectivity is negative. One such case is where what's intended is *projective illusion* or "projection" in the Freudian sense. Now projection is actually a pathological condition, to the extent that it represents a mistake, an illusion based on a sort of confusion between a characteristic of oneself (which isn't acknowledged) and which is ascribed to others though it isn't actually there. But there are closely related phenomena that carry no such stigma. Indeed, some people have suggested that simulation (a sort of systematic but non-self-deceptive projection) plays a crucial role in our understanding of other people. (Gordon 1986, 1992; Goldman 1992; contra, see Stich and Nichols 1992).[13]

11. SEEING-AS

Among the subtleties of my perceptual point of view and of my experiences (to which I return in the next section), lies a facet of subjectivity which does not appear to be exhausted by the previous descriptions. This is the fact that what I perceive, I commonly *perceive-as-something-or-other*. When I see a duck-rabbit, how many qualia do I see? In the paper cited above, Ramachandran and Hirstein go so far as to propose the following unabashedly functional hypothesis: when I look at a duck-rabbit figure, I can "see-it-as" only one (at least one at a time), because that's precisely among the *functions* of consciousness: to fix "irrevocably" what we see so as to make it possible for us to undertake unambiguous action. Though Hamlet says conscience makes cowards of us all, yet it is *consciousness*, on this hypothesis, that has the job of keeping us from the Hamlet syndrome.

To see something as a so-and-so, is to see it in a way that involves intentionality. Yet some sensations are sometimes held to be experienced non-intentionally: the typical example is pain.

Pains, however, are *sensed as painful*. Indeed, (Kripke 1980) used this fact to revive an old argument against the identity theory of sensations and brain states. That classical argument went as follows:

A painful sensation S is not just painful, but *essentially painful*. Any neural process N with which we might claim to identify it, however, could only be contingently painful. Therefore, by Leibniz's law, N couldn't be identical with S.

In order to appreciate the modifications that Kripke brought to this argument, recall that he loosened the bonds linking the analytic, the a priori, and the necessary, and their contraries, the synthetic, the a posteriori and the contingent. His analysis requires us to appreciate that some statements, such as the identity statement 'Water is H_2O', or 'heat is mean molecular momentum', might be both *necessary* and *a posteriori*. The necessity involved is ontological, not epistemic:

Kripke allows that there is nothing wrong with the statement that for all we knew there might have been some different physical process responsible for the feeling of heat. But as it happens, there isn't and there couldn't be compatibly with the laws of nature (Kripke 1980, 333).

Kripke's doctrine actually constitutes a minor puzzle in the history of contemporary philosophy. For having thus relaxed the link between what can be known a priori and what is necessary opens the door to an identity theory immune to the old arguments about what can be imagined to be identical. Yet no sooner did Kripke open this door than he tried to push it shut: nothing is more certain, he argued, than the fact that to be painful is a necessary property of any pain. So if N is really identical to S, then it must be the case that N is necessarily painful too. And that Kripke finds incredible, on the ground that while in the case of heat there is something, namely the process of molecular motion, between the sensation and the heat, there is no analogue to this "something in the middle" in the case of pain. For the *essence* of pain is nothing but quale felt by the sufferer. (339)

But what's wrong with the other possibility? Why not say that the neural process N is indeed necessarily painful? Since Kripke has duly shown that some necessary truths can be known a posteriori, the burden of proof is now on him to show that this isn't one such necessary truth. All that we need grant his argument is the analyticity of 'pain is painful', and the fact that any physiological state's painfulness, by contrast, is synthetic. Given that, however, it remains perfectly possible that some physiological state really is *necessarily* painful just as water is necessarily H_2O, regardless of the fact that neither, in advance of scientific knowledge, can be expected to seem analytic.

12. PHENOMENAL EXPERIENCE

I have left till last the most hotly contested of subjectivity's battlefields. Chalmers puts the centrality of qualia in these terms:

> The problem of explaining these philosophical qualities is just the problem of consciousness. This is the really hard part of the mind-body problem. (Chalmers 1996, 4)

But it seems to be characteristic of those who take this form of subjectivity as central that any attempt at *explaining* our talk about qualia in materialist term is taken as a *refusal to take the hard problem seriously*. Dennett, for example, has repeatedly been accused of denying that we are conscious: "Dennett thinks there are no such things as qualia, subjective experiences, first-person phenomena, or any of the rest of it." ((Searle 1997, 99) Apart from being admirably unambiguous about the "charge" against Dennett, this is a nice example of how the vast and unspecified "rest of it" is thrown into the same bag as qualia. Dennett has repeatedly denied denying that we have conscious experience, but since he has indeed also "quined qualia" (Dennett 1990), it's clear without assessing the argument that Searle's mere assertion of the intuition that qualia just are the "data" doesn't settle the matter. Dennett doesn't deny that we are conscious, or that we have experiences, "and the rest of it." He just claims that the philosophical mystery made about qualia can be dispelled once it is kept separate from other theoretical issues that surround it. He has also argued that once one focuses on the *functions* of qualia — which can perfectly well be discussed in the third person — there is nothing left for the ineffably private qualia to be.

The issue of qualia seems to me open to a *vice-grip strategy*, which consists in squeezing the irreducibility of qualia between two complementary poles, the materialist identity theory and functionalism. Both are reductive in the sense that they propose third-person accounts of the qualia in question.

(1) *Materialism*: Churchland has recently argued that the identity theory of qualia can be rehabilitated. His strategy is illustrated by showing how the colour solid is isomorphic to the solid generated by the 3-dimensional structure of the antagonistic receptors which receive input from the three types of cones. The essential strategy here consists in the challenge: "what more do you want than full formal coherence between the physiological mechanisms and the phenomenological structure of the colour solid (or *mutatis mutandis*)?" (Churchland and Churchland 1998). If it is then objected that *correlations don't establish identity*, the objector owes an account of what more is required. The answer can only be that two objects, however perfectly correlated, might differ in their causes and effects. The argument is then ready to be turned over to the functionalist.

(2) *Functionalism*: This is also best summed up in a challenge, the *zombie challenge*: if you can imagine some being whose reactions to a given scene (sound, sight, stimulus, or whatever) are like yours in every possible way, including synesthetic, associative, recollections evoked, etc.) can you really also imagine that this being might *not* be like you merely in lacking qualitative experience? If you can, then subjective consciousness, as such, is strictly epiphenomenal in a sense so strong as to make it "a concept that has no utility whatsoever" (Dennett 1991, 402).

At this point, an objector might suggest that functional equivalence is not enough, since two functionally equivalent items might be substantially different. But this objector is one that can safely be left in the hands of the identity theorist, who can once again appeal to the structural correlations between qualia and their physiological underpinnings.

At the risk of dismissing this huge debate with cavalierly short shrift, this vice-grip strategy seems to me sufficiently promising to suggest that the subjectivity of qualia does not actually constitute an insolubly "hard problem".

CONCLUSION

Some of these "senses" or "aspects" of subjectivity may be redundant. I certainly am not confident that none could, by means of some ingenious argument, be reduced or assimilated to others. But to establish conclusively that they are all distinct would involve 11! or 65 pairwise comparison: I leave this as a rich mine of thesis topics for future doctoral students to explore. All I will rest on now is the thought that there are plausible, non-mystifying avenues of research open on each of the twelve forms of subjectivity I have described, and that in such piecemeal solutions lies the hope of solving the so-called problem of subjective consciousness.

Ronald de Sousa
University of Toronto
Canada

NOTES

1. Lycan's own strategy, however, is much like the one I espouse, though he concentrates on anatomizing the term 'consciousness' rather than the term 'subjectivity'. See Lycan (1996, 2-7).
2. See “Contre la phénoménologie”, forthcoming, available at http://www.chass.utoronto.ca/sousa/contrephen.html.
3. Christians solved this problem with the usual combination of ingenuity and absurdity: God, while eternal, becomes incarnate, just to show that it's not logically impossible for Him to say things like "Thank Goodness it's Friday."
4. Nagel, in *The View from Nowhere*, conflates several types of subjectivity, though I haven't ascertained that all twelve can be found there. Not surprisingly, subjectivity turns out to be something of a mystery, and an irreducible one at that.
5. Note how the language here supports the idea that we are dealing with a single and mysterious "ontological mode": by speaking of a "consequence" of subjectivity, rather than a component, type, or sense of subjectivity, Searle suggests that we can have a prior knowledge of this "ontological mode" and that we are merely drawing out its implications.
6. The dilemma I have just articulated leaves out what may appear to be the most poignant case: namely where I am already too gaga to express any opinion. In those cases, however, we might justify giving a "living will" authority by default even in a Parfitean world, as a sort of legal fiction, just as we grant authority by default to nearest relatives for certain other decisions affecting the welfare of the incompetent.
7. By the existentialists, by Stuart Hampshire in his *Thought and Action* (Hampshire 1983) and by Charles Taylor, in *Sources of the Self* (Taylor 1989) among others.
8. Descartes's *cogito* can also be credited with stressing subjectivity in a sense I'm not sure how to classify. This emerges from the consideration that if we try to paraphrase his argument in *Med. II* as a syllogism with a universal major premise, “Whatever thinks, exists”: the argument will collapse because the premise is false, since thinking is done admirably well by many a fictional character.

9. In the phenomenological literature, there is supposed to be a "second phenomenological reduction" that has to do with ownness; but since I have never understood what that is, I confine myself to the non-phenomenological literature.
10. Cf. Aristotle, 1963 *Categories* 2:1A25-8 "There is, lastly, a class of things which are neither present in a subject nor predicable of a subject, such as the individual man or the individual horse. But, to speak more generally, that which is individual and has the character of a unit is never predicable of a subject. Yet in some cases there is nothing to prevent such being present in a subject. Thus a certain point of grammatical knowledge is present in a subject."
11. One set of thought experiments are attributed to Zuboff in Tye (1995, 78 ff). Ramachandran and Hirstein (1997) offer similar speculations, in which judicious rewiring of brain connections results in two people sharing or exchanging experiences. They infer that sensations are actually not in principle unobservable by others. The contrary appearance, they argue, is due simply to the fact that in ordinary situations I can only have access to the experiences of others through "translation". But if my brain were wired in just the right way to yours I would have direct access to your mental states.
12. A novel the name of which I've forgotten tells the story of a detective who suffers from amnesia and gradually comes to the conclusion that the criminal he is tracking is *he himself.*
13. For further debate see (Davies and Stone 1995).

REFERENCES

Ainslie, G. (1992). *Psicoeconomics: The Strategic Interaction of Successive Motivational States Within the Person*. Cambridge: Cambridge University Press.

Aristotle. (1963). *Categories and de Interpretatione* (J. Ackrill, trans and notes). Oxford: Oxford University Press.

Berkeley, G. (1957). *A Treatise Concerning the Principles of Human Knowledge* (C. M. Turbayne, ed & introd by). Liberal Arts Press. Indianapolis, New York: Bobbs-Merril.

Campbell, S. (1998). *Interpreting the Personal: Expression and the Formation of Feeling*. Ithaca: Cornell University.

Castaneda, H.-N. (1988). Self-consciousness, demonstrative reference, and self-ascription. In J. E. Tomberlin (Ed.), *Philosophical Perspectives*, pp. 405-454. Atascadero: Ridgeview.

Chalmers, D. J. (1996). *The Conscious Mind: In Search of a Fundamental Theory*. Oxford; New York: Oxford University Press.

Churchland, P. M. & Churchland, P. (1998). *On the Contrary: Critical Essays*. Cambridge, MA: MIT Press.

Cole, J. (1997). On 'being faceless': Selfhood and facial embodiment. *Journal of Consciousness Studies,* **4**(5-6), 467-484.

Cole, J. (1998). About Face. Cambridge, MA : MIT Press A Bradford Book.

Davies, M. & Stone, T. (eds). (1995). Mental Simulation: Evaluations and Applications. *Readings in Mind and Language*. Oxford: Blackwell.

Dennett, D. (1990). Quining qualia. In *Mind and Cognition: A Reader*, pp. 519-547. Oxford : Blackwell.

Dennett, D. C. (1991). *Consciousness Explained.* Boston, Toronto, London: Little, Brown.

Descartes, R. (1984-85) [1649]. *The Philosophical Writings of Descartes*. (J. Cottingham, R. Stoothoff & D. Murdoch, Trans.). Cambridge: Cambridge University Press.

Dretske, F. (1995). *Naturalizing the Mind.* Cambridge, MA: MIT Press.

Goldman, A. I. (1992). In defense of the simulation theory. *Mind and Language*, **7**, 104-119.

Gopnik, A. (1993) How we know our own minds: the illusion of first-person knowledge of intentionality. *Behavioral and Brain Sciences*, **16**, 145-171.

Gordon, R. M. (1986). Folk psychology and simulation. *Mind and Language*, **1**, 158-171.

Gordon, R. M. (1992). The simulation theory: Objections and misconceptions. *Mind and Language*, **7**, 11-34.

Hampshire, S. (1983). *Thought and Action*, 2nd ed. Notre Dame, Indiana: University of Notre Dame Press.

Johnson-Laird, P. N. (1988). *The Computer and the Mind: An Introduction to Cognitive Science*. Cambridge, MA: Harvard University Press.

Kripke, S. A. (1980). *Naming and Necessity*. Cambridge, MA: Harvard University Press.

Lycan, W. (1996). *Consciousness and Experience*. Cambridge, MA: MIT Press.

Meltzoff, A. & Gopnik, A. (1993). The role of imitation in understanding persons and developing a theory of mind. In S. Baron-Cohen, H. Tager-Flusberg & D. Cohen (eds), *Understanding Other Minds: Perspectives from Autism*. Oxford: Oxford University Press.

Moravia, S. (1995) [1986]. *The Enigma of the Mind, Tr. from the Italian L'enigma della mente.* Cambridge: Cambridge University Press.

Nagel, T. (1986). *The View from Nowhere*. Oxford: Oxford University Press.

Parfit, D. (1971). *Personal identity*. Philosophical Review, **80**.

Parfit, D. (1984). *Reasons and Persons*. Oxford: Oxford University Press .

Perry, J. (1979). The problem of the essential indexical. *Nous*, **3**, 3-21.

Proust, J. (1997) *Comment l'esprit vient aux bêtes: essai sur la représentation*. Paris: NRF Essais.

Ramachandran, V. & Hirstein, W. (1997). Three laws of qualia: What neurology tells us about the biological functions of consciousness. *Journal of Consciousness Studies*, **4**(5-6), 429-457.

Searle, J. R. (1992). *The Rediscovery of the Mind*. Cambridge, MA : MIT Press. A Bradford Book.

Searle, J. R. (1997). *The Mystery of Consciousness, and Exchanges with Daniel C. Dennett and David J. Chalmers.* London: Granta.

Sellars, W. (1963). *Science, Perception and Reality.* New York: Humanities Press.

Sober, E. & Wilson, D. S. (1998). *Unto Others: The Evolution and Psychology of Unselfish Behavior.* Cambridge, MA: Harvard University Press.

Stich, S. & Nichols, S. (1992). Folk psychology: Simulation or tacit theory? *Mind and Language*, **7**, 35-71.

Strawson, G. (1997). The self. *Journal of Consciousness Studies*, **4**(5/6), 405-428.

Taylor, C. (1989). *Sources of the Self.* Cambridge, MA: Harvard University Press.

Thompson, E. (1995). *Colour vision: A study in cognitive science and the philosophy of Perception.* London: Routledge.

Tye, M. (1990). A representational theory of pains and their phenomenal character. *Philosophical Perspectives,* **9**, 223-239, J. Tomberlin (Ed.). Atascadero: Ridgeview Publishing.

Tye, M. (1995). *Ten Problems of Consciousness.* Cambridge, MA: MIT Press.

NAME INDEX

SUBJECT INDEX

PHILOSOPHICAL STUDIES SERIES

1. Jay F. Rosenberg: *Linguistic Representation.* 1974 ISBN 90-277-0533-X
2. Wilfrid Sellars: *Essays in Philosophy and Its History.* 1974 ISBN 90-277-0526-7
3. Dickinson S. Miller: *Philosophical Analysis and Human Welfare.* Selected Essays and Chapters from Six Decades. Edited with an Introduction by Lloyd D. Easton. 1975 ISBN 90-277-0566-6
4. Keith Lehrer (ed.): *Analysis and Metaphysics.* Essays in Honor of R. M Chisholm. 1975 ISBN 90-277-0571-2
5. Carl Ginet: *Knowledge, Perception, and Memory.* 1975 ISBN 90-277-0574-7
6. Peter H. Hare and Edward H. Madden: *Causing, Perceiving and Believing.* An Examination of the Philosophy of C. J. Ducasse. 1975 ISBN 90-277-0563-1
7. Hector-Neri Castañeda: *Thinking and Doing.* The Philosophical Foundations of Institutions. 1975 ISBN 90-277-0610-7
8. John L. Pollock: *Subjunctive Reasoning.* 1976 ISBN 90-277-0701-4
9. Bruce Aune: *Reason and Action.* 1977 ISBN 90-277-0805-3
10. George Schlesinger: *Religion and Scientific Method.* 1977 ISBN 90-277-0815-0
11. Yirmiahu Yovel (ed.): *Philosophy of History and Action.* Papers presented at the First Jerusalem Philosophical Encounter (December 1974). 1978 ISBN 90-277-0890-8
12. Joseph C. Pitt (ed.): *The Philosophy of Wilfrid Sellars: Queries and Extensions.* 1978 ISBN 90-277-0903-3
13. Alvin I. Goldman and Jaegwon Kim (eds.): *Values and Morals.* Essays in Honor of William Frankena, Charles Stevenson, and Richard Brandt. 1978 ISBN 90-277-0914-9
14. Michael J. Loux: *Substance and Attribute.* A Study in Ontology. 1978 ISBN 90-277-0926-2
15. Ernest Sosa (ed.): *The Philosophy of Nicholas Rescher.* Discussion and Replies. 1979 ISBN 90-277-0962-9
16. Jeffrie G. Murphy: *Retribution, Justice, and Therapy.* Essays in the Philosophy of Law. 1979 ISBN 90-277-0998-X
17. George S. Pappas (ed.): *Justification and Knowledge.* New Studies in Epistemology. 1979 ISBN 90-277-1023-6
18. James W. Cornman: *Skepticism, Justification, and Explanation.* With a Bibliographic Essay by Walter N. Gregory. 1980 ISBN 90-277-1041-4
19. Peter van Inwagen (ed.): *Time and Cause.* Essays presented to Richard Taylor. 1980 ISBN 90-277-1048-1
20. Donald Nute: *Topics in Conditional Logic.* 1980 ISBN 90-277-1049-X
21. Risto Hilpinen (ed.): *Rationality in Science.* Studies in the Foundations of Science and Ethics. 1980 ISBN 90-277-1112-7
22. Georges Dicker: *Perceptual Knowledge.* An Analytical and Historical Study. 1980 ISBN 90-277-1130-5
23. Jay F. Rosenberg: *One World and Our Knowledge of It.* The Problematic of Realism in Post-Kantian Perspective. 1980 ISBN 90-277-1136-4
24. Keith Lehrer and Carl Wagner: *Rational Consensus in Science and Society.* A Philosophical and Mathematical Study. 1981 ISBN 90-277-1306-5
25. David O'Connor: *The Metaphysics of G. E. Moore.* 1982 ISBN 90-277-1352-9

PHILOSOPHICAL STUDIES SERIES

26. John D. Hodson: *The Ethics of Legal Coercion.* 1983 ISBN 90-277-1494-0
27. Robert J. Richman: *God, Free Will, and Morality.* Prolegomena to a Theory of Practical Reasoning. 1983 ISBN 90-277-1548-3
28. Terence Penelhum: *God and Skepticism.* A Study in Skepticism and Fideism. 1983 ISBN 90-277-1550-5
29. James Bogen and James E. McGuire (eds.): *How Things Are.* Studies in Predication and the History of Philosophy of Science. 1985 ISBN 90-277-1583-1
30. Clement Dore: *Theism.* 1984 ISBN 90-277-1683-8
31. Thomas L. Carson: *The Status of Morality.* 1984 ISBN 90-277-1619-9
32. Michael J. White: *Agency and Integrality.* Philosophical Themes in the Ancient Discussions of Determinism and Responsibility. 1985 ISBN 90-277-1968-3
33. Donald F. Gustafson: *Intention and Agency.* 1986 ISBN 90-277-2009-6
34. Paul K. Moser: *Empirical Justification.* 1985 ISBN 90-277-2041-X
35. Fred Feldman: *Doing the Best We Can.* An Essay in Informal Deontic Logic. 1986 ISBN 90-277-2164-5
36. G. W. Fitch: *Naming and Believing.* 1987 ISBN 90-277-2349-4
37. Terry Penner: *The Ascent from Nominalism.* Some Existence Arguments in Plato's Middle Dialogues. 1987 ISBN 90-277-2427-X
38. Robert G. Meyers: *The Likelihood of Knowledge.* 1988 ISBN 90-277-2671-X
39. David F. Austin (ed.): *Philosophical Analysis.* A Defense by Example. 1988 ISBN 90-277-2674-4
40. Stuart Silvers (ed.): *Rerepresentation.* Essays in the Philosophy of Mental Representation. 1988 ISBN 0-7923-0045-9
41. Michael P. Levine: *Hume and the Problem of Miracles.* A Solution. 1989 ISBN 0-7923-0043-2
42. Melvin Dalgarno and Eric Matthews (eds.): *The Philosophy of Thomas Reid.* 1989 ISBN 0-7923-0190-0
43. Kenneth R. Westphal: *Hegel's Epistemological Realism.* A Study of the Aim and Method of Hegel's *Phenomenology of Spirit.* 1989 ISBN 0-7923-0193-5
44. John W. Bender (ed.): *The Current State of the Coherence Theory.* Critical Essays on the Epistemic Theories of Keith Lehrer and Laurence BonJour, with Replies. 1989 ISBN 0-7923-0220-6
45. Roger D. Gallie: *Thomas Reid and 'The Way of Ideas'.* 1989 ISBN 0-7923-0390-3
46. J-C. Smith (ed.): *Historical Foundations of Cognitive Science.* 1990 ISBN 0-7923-0451-9
47. John Heil (ed.): *Cause, Mind, and Reality.* Essays Honoring C. B. Martin. 1989 ISBN 0-7923-0462-4
48. Michael D. Roth and Glenn Ross (eds.): *Doubting.* Contemporary Perspectives on Skepticism. 1990 ISBN 0-7923-0576-0
49. Rod Bertolet: *What is Said.* A Theory of Indirect Speech Reports. 1990 ISBN 0-7923-0792-5
50. Bruce Russell (ed.): *Freedom, Rights and Pornography.* A Collection of Papers by Fred R. Berger. 1991 ISBN 0-7923-1034-9
51. Kevin Mulligan (ed.): *Language, Truth and Ontology.* 1992 ISBN 0-7923-1509-X

PHILOSOPHICAL STUDIES SERIES

52. Jesús Ezquerro and Jesús M. Larrazabal (eds.): *Cognition, Semantics and Philosophy*. Proceedings of the First International Colloquium on Cognitive Science. 1992 ISBN 0-7923-1538-3
53. O.H. Green: *The Emotions*. A Philosophical Theory. 1992 ISBN 0-7923-1549-9
54. Jeffrie G. Murphy: *Retribution Reconsidered*. More Essays in the Philosophy of Law. 1992 ISBN 0-7923-1815-3
55. Phillip Montague: *In the Interests of Others*. An Essay in Moral Philosophy. 1992 ISBN 0-7923-1856-0
56. Jacques-Paul Dubucs (ed.): *Philosophy of Probability*. 1993 ISBN 0-7923-2385-8
57. Gary S. Rosenkrantz: *Haecceity*. An Ontological Essay. 1993 ISBN 0-7923-2438-2
58. Charles Landesman: *The Eye and the Mind*. Reflections on Perception and the Problem of Knowledge. 1994 ISBN 0-7923-2586-9
59. Paul Weingartner (ed.): *Scientific and Religious Belief*. 1994 ISBN 0-7923-2595-8
60. Michaelis Michael and John O'Leary-Hawthorne (eds.): *Philosophy in Mind*. The Place of Philosophy in the Study of Mind. 1994 ISBN 0-7923-3143-5
61. William H. Shaw: *Moore on Right and Wrong*. The Normative Ethics of G.E. Moore. 1995 ISBN 0-7923-3223-7
62. T.A. Blackson: *Inquiry, Forms, and Substances*. A Study in Plato's Metaphysics and Epistemology. 1995 ISBN 0-7923-3275-X
63. Debra Nails: *Agora, Academy, and the Conduct of Philosophy*. 1995 ISBN 0-7923-3543-0
64. Warren Shibles: *Emotion in Aesthetics*. 1995 ISBN 0-7923-3618-6
65. John Biro and Petr Kotatko (eds.): *Frege: Sense and Reference One Hundred Years Later*. 1995 ISBN 0-7923-3795-6
66. Mary Gore Forrester: *Persons, Animals, and Fetuses*. An Essay in Practical Ethics. 1996 ISBN 0-7923-3918-5
67. K. Lehrer, B.J. Lum, B.A. Slichta and N.D. Smith (eds.): *Knowledge, Teaching and Wisdom*. 1996 ISBN 0-7923-3980-0
68. Herbert Granger: *Aristotle's Idea of the Soul*. 1996 ISBN 0-7923-4033-7
69. Andy Clark, Jesús Ezquerro and Jesús M. Larrazabal (eds.): *Philosophy and Cognitive Science: Categories, Consciousness, and Reasoning*. Proceedings of the Second International Colloquium on Cogitive Science. 1996 ISBN 0-7923-4068-X
70. J. Mendola: *Human Thought*. 1997 ISBN 0-7923-4401-4
71. J. Wright: *Realism and Explanatory Priority*. 1997 ISBN 0-7923-4484-7
72. X. Arrazola, K. Korta and F.J. Pelletier (eds.): *Discourse, Interaction and Communication*. Proceedings of the Fourth International Colloquium on Cognitive Science. 1998 ISBN 0-7923-4952-0
73. E. Morscher, O. Neumaier and P. Simons (eds.): *Applied Ethics in a Troubled World*. 1998 ISBN 0-7923-4965-2
74. R.O. Savage: *Real Alternatives, Leibniz's Metaphysics of Choice*. 1998 ISBN 0-7923-5057-X
75. Q. Gibson: *The Existence Principle*. 1998 ISBN 0-7923-5188-6
76. F. Orilia and W.J. Rapaport (eds.): *Thought, Language, and Ontology*. 1998 ISBN 0-7923-5197-5

PHILOSOPHICAL STUDIES SERIES

77. J. Bransen and S.E. Cuypers (eds.): *Human Action, Deliberation and Causation*. 1998 ISBN 0-7923-5204-1
78. R.D. Gallie: *Thomas Reid: Ethics, Aesthetics and the Anatomy of the Self*. 1998 ISBN 0-7923-5241-6
79. K. Korta, E. Sosa and X. Arrazola (eds.): *Cognition, Agency and Rationality*. Proceedings of the Fifth International Colloquium on Cognitive Science. 1999 ISBN 0-7923-5973-9
80. M. Paul: *Success in Referential Communication*. 1999 ISBN 0-7923-5974-7
81. E. Fischer: *Linguistic Creativity*. Exercises in 'Philosophical Therapy'. 2000 ISBN 0-7923-6124-5
82. R. Tuomela: *Cooperation*. A Philosophical Study. 2000 ISBN 0-7923-6201-2
83. P. Engel (ed.): *Believing and Accepting*. 2000 ISBN 0-7923-6238-1
84. W.L. Craig: *Time and the Metaphysics of Relativity*. 2000 ISBN 0-7923-6668-9
85. D.A. Habibi: *John Stuart Mill and the Ethic of Human Growth*. 2001 ISBN 0-7923-6854-1
86. M. Slors: *The Diachronic Mind*. An Essaý on Personal Identity, Psychological Continuity and the Mind-Body Problem. 2001 ISBN 0-7923-6978-5
87. L.N. Oaklander (ed.): *The Importance of Time*. Proceedings of the Philosophy of Time Society, 1995–2000. 2001 ISBN 1-4020-0062-6
88. M. Watkins: *Rediscovering Colors*. A Study in Pollyanna Realism. 2002 ISBN 1-4020-0737-X
89. W.F. Vallicella: *A Paradigm Theory of Existence*. Onto–Theology Vindicated. 2002 ISBN 1-4020-0887-2
90. M. Hulswit: *From Cause to Causation*. A Peircean Perspective. 2002 ISBN 1-4020-0976-3; Pb 1-4020-0977-1
91. D. Jacquette (ed.): *Philosophy, Psychology, and Psychologism*. Critical and Historical Readings on the Psychological Turn in Philosophy. 2003 ISBN 1-4020-1337-X
92. G. Preyer, G. Peter and M. Ulkan (eds.): *Concepts of Meaning*. Framing an Integrated Theory of Linguistic Behavior. 2003 ISBN 1-4020-1329-9
93. W. de Muijnck: *Dependencies, Connections, and Other Relations*. A Theory of Mental Causation. 2003 ISBN 1-4020-1391-4
94. N. Milkov: *A Hundred Years of English Philosophy*. 2003 ISBN 1-4020-1432-5
95. E.J. Olsson (ed.): *The Epistomology of Keith Lehrer*. 2003 ISBN 1-4020-1605-0
96. D.S. Clarke: *Sign Levels*. Language and Its Evolutionary Antecedents. 2003 ISBN 1-4020-1650-6
97. A. Meirav: *Wholes, Sums and Unities*. 2003 ISBN 1-4020-1660-3
98. C.H. Conn: *Locke on Esence and Identity*. 2003 ISBN 1-4020-1670-0
99. J.M. Larrazabal and L.A. Pérez Miranda: *Language, Knowledge, and Representation. Proceedings of the Sixth International Colloquium on Cognitive Science (ICCS-99)*. 2004 ISBN 1-4020-2057-0
100. P. Ziff: *Moralities*. A Diachronic Evolution Approach. 2004 ISBN 1-4020-1891-6
101. J.A. Corlett: Terorism: *A Philosophical Analysis*. 2003 ISBN 1-4020-1694-8; Pb 1-4020-1695-6

PHILOSOPHICAL STUDIES SERIES

102. K. Korta and J.M. Larrazabal (eds.): *Truth, Rationality, Cognition, and Music*. Proceedings of the Seventh International Colloquium on Cognitive Science. 2003 ISBN 1-4020-1912-2

KLUWER ACADEMIC PUBLISHERS – DORDRECHT / BOSTON / LONDON